Naturalists' Handbooks 5

Hoverflies

FRANCIS S. GILBERT
Department of Zoology,
Nottingham University

With illustrations by Steven J. Falk

Revised edition

Pelagic Publishing
www.pelagicpublishing.com

Published by Pelagic Publishing
www.pelagicpublishing.com
PO Box 725, Exeter, EX1 9QU, UK

Hoverflies
Naturalists' Handbooks 5

Series editors
S. A. Corbet and R. H. L. Disney

ISBN 978-1-907807-59-6 Paperback

Digital reprint edition of:
ISBN 0-85546-255-8 (1993) Paperback
ISBN 0-85546-256-6 (1993) Hardback

This revised 2nd edition:
Text © Pelagic Publishing 2015

Illustrations © Steven J. Falk

British Library Cataloguing in Publication Data
A catalogue record for this book is available from the British Library.

Contents

Editors' preface

Many people without a university training in biology have the opportunity and inclination to study natural history but lack the knowledge to do so in a confident and productive way. The books in this series offer them the information and ideas needed to plan an investigation, and the practical guidance needed to carry it out. They draw attention to regions on the frontiers of current knowledge where amateur studies have much to offer. We hope the readers will derive as much satisfaction from their biological explorations as we have done. Even in Britain, the identification of many groups remains a barrier to ecological research, and it is an important feature of the series that it enables readers to name the organisms they work with. Other books in the series do this by means of keys. Hoverflies are well served by the comprehensive, well-illustrated and user-friendly keys in British Hoverflies by A. E. Stubbs and S. J. Falk, so instead of keys this book offers illustrations and descriptions characterising selected species common in Britain. We thank the Natural Environment Research Council for a grant towards the cost of the colour illustrations (which originally appeared in Stubbs and Falk's British Hoverflies) and the line drawings.

S.A.C.

R.H.L.D

Acknowledgements

I would like to thank Sally Corbet for introducing me to hoverflies, and for help and encouragement during my studies; Graham Rotheray for his enthusiasm, and particularly for telling me about larval biology; Ivan Perry for advice on identification and biology of hoverflies; and Oliver Prŷs-Jones for exploring floral ecology with me and for correcting many of my erroneous ecological ideas. I am pleased to be able to express my gratitude to the Natural Environment Research Council, Gonville & Caius College, Cambridge, The Commonwealth Fund of New York, and especially St John's College, Cambridge, for their logistic and financial support. This book could not have been written without the support of my wife, Hilary, who edited several early drafts. Later versions were greatly improved by the comments of Graham Rotheray, Peter Stiling, Dennis Unwin, Sally Corbet, Henry Disney, Bill Hawley, Ivan Perry, Martin Speight, Steven Falk and Alan Stubbs.

I would like to thank Sally Corbet for inviting me to write the first edition of this book, and affording me so much pleasure in doing so. I am even more pleased at having had the opportunity of revising it in this second edition, since I have been able to correct most of the mistakes, and add some of the more interesting discoveries of the last few years. I dedicated the first edition to G. Clifford Evans, and the Fellows of St John's College, Cambridge, in gratitude for the six marvellous years I spent in the College. I dedicate the second edition to my friend and colleague, for the pleasure of our joint work on hoverflies together.

To Graham Rotheray

F.S.G.

January 1993

1 Introduction

Fig. 1. A typical hoverfly

Fig. 2. Thorax of *Episyrphus balteatus*.

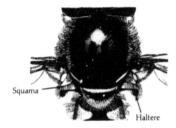

Squama

Haltere

Almost everyone will have seen hoverflies, and perhaps wondered what they are, and how they live. Many people will have been struck by their resemblance to bees or wasps, and probably many will have mistaken them for these groups. This book should enable anyone to identify the commonest hoverflies to be found in the British Isles, and to carry out much-needed research into their ecology and biology. Anyone sufficiently interested in finding out more about these insects will be able to make valuable contributions to our knowledge, for most of the research that will suggest itself to you as you read will need only the most basic of equipment, patience and a notepad. It is not necessary to be at a university to carry out ecological research that leads to publishable results.

Being insects, hoverflies have six legs and a body divided into three parts: the head, thorax and abdomen. They are true flies (Order Diptera): attached to the thorax they have one pair of wings, and a pair of club-shaped organs called halteres (fig. 2) that act as gyroscopes to control flight movements. Many families within the Diptera can be recognised from the pattern of veins in the wings, and hoverflies are no exception; fig. 3 shows the general plan of a hoverfly's wing. Distinguishing features are the 'false vein' in the middle of the wing, the very long cell 'C', and the fact that the vein running parallel to the outer edge of the wing (vein a in fig. 3) closes the cells 'A' and 'B'. Look at the wing veins in the colour plates, and compare them with the similar veins of another fly, the housefly (fig. 4). This will demonstrate what to look for; the housefly has no 'false vein', and cell 'A' is open (just) to the wing margin. Once a few hoverflies have been identified and their behaviour observed, you will soon get a 'feel' for which are hoverflies and which are not. You will then be able to

Fig. 3. Wing veins of a typical hoverfly.

Fig. 4. Wing veins of a housefly.

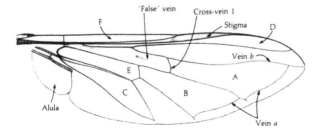

recognise them on sight from many metres away, without having to catch them to look at their wings.

Hoverflies can be found almost anywhere. The adults feed mostly on nectar and pollen from flowers, but can also be seen feeding from leaf surfaces, usually where aphids have produced a film of honeydew on the leaves. A garden with flowers, especially of the family Compositae (e.g. asters or Michaelmas daisies), will attract many species even in large cities. Most British hoverflies are naturally associated with woodland habitats, but many have adapted to environments provided by man, particularly parks and gardens. The tremendous diversity of hoverflies to be found in a suburban garden has been described by Owen (1991)*, who recorded 93 species during the course of 12 years. Hoverflies can be caught from early spring to late autumn; the rarer species are often on the wing in spring or early summer, but most common species appear during late July and August.

The larvae can be found throughout the year in a wide variety of habitats. Often the most fruitful searches are under fallen leaves in late autumn or during the winter. Larvae found at this time are often overwintering in a type of 'suspended animation' called diapause. This book does not deal with the identification of larvae: if you wish to do this, you should consult Rotheray's (1989a) book in this series, or look at the references to larvae listed on p. 63. The best way of identifying larvae, however, is to allow them to develop into adults!

Forty-two species of hoverfly are included in this book, covering all the common species of the British Isles, though some 250 species in total have been recorded here. Few have common names, and therefore their internationally recognised Latin (or Latinised Greek) names have to be used. In the case of *Platycheirus albimanus*, for example,'*Platycheirus*' (Greek: 'flat-hands') refers to a genus (plural: genera) to which several closely related species belong; '*albimanus*' (Latin: 'white-handed') tells you which species it is. At the first mention of a species in a publication, the name of the person who gave the species its name is also included (usually abbreviated to a few letters). Thus the full name would be *Platycheirus albimanus* (Fabricius). These Latin names are convenient: you can tell instantly which species are closely related because they are in the same genus, and therefore have the same generic name. Each genus may be closely or distantly related to other genera, and this is recognised by grouping related genera into tribes, and related tribes into subfamilies. There are three subfamilies in the family Syrphidae, the hoverflies.

* References cited under the author's name in the text appear in full in References and further reading on p. 63

A classification of species described in this book is shown in table 1. Only two of the three subfamilies are included; the third, the Microdontinae, is a rather rare group in Britain.

The common species can nearly all be identified from their colour patterns, by matching a specimen to the colour plates. However, to be absolutely sure that a correct match has been found, you must check using the descriptions in the text, as the colour patterns can be variable. Dark, or even black forms are frequent, particularly in the genera *Platycheirus* and *Melanostoma*. The patterns cannot therefore be relied upon to name specimens invariably. In addition, several species are indistinguishable from one another at first sight, and close examination is necessary. As some important features involve, for example, the distribution of hairs over various parts of the body, a binocular microscope with adequate lighting will be necessary. Most schools have them, or could arrange for you to use one. Good lighting is very important: you will need a focussed light to see *all* the characters mentioned, although nearly all the species should be identifiable without specialised light sources. Measurement of small objects such as eggs requires an ocular micrometer, a disc of glass with a graduated scale marked on it that fits into the eyepiece of the microscope. With good viewing equipment, even the smallest and most nondescript of insects becomes a fantastically coloured, armour-plated creature from an alien world. Such close observation and dissection of insects reveals the extraordinary intricacies of their construction.

We hope that this book will encourage you to study these delightful creatures; such study is not only rewarding in its own right, but yields results that can be published, and that therefore contribute to our understanding of the natural world. Some of the suggested techniques require unusual pieces of equipment, but access to these should not prove difficult. The book consists of three main chapters, two concerning the biology of hoverflies, and one containing descriptions of the species. Techniques referred to in the text are described in chapter 5. Finally there is a list of selected references for those who wish to complement their studies with background material. The reference list also contains books that provide background material to the study of ecology in general, as an aid to thinking about ecological experiments and aims.

It is often claimed by non-scientists that those who investigate just one or a group of species will rapidly discover everything about those species. Even a brief acquaintance with the complexity of ecology, and with the methods used to unravel some of the problems, provokes the opposite reaction: how on earth can we ever hope really to understand what is going on? Building upon past studies, the scope of ecological questions is limitless, as new theories are proposed and new data are required to test them. But it is not necessary to have worked out in detail the entire biology of one species in order to publish the results of a study. Each complete topic may be published separately. Each of the papers involving experimentation published in, say, the *Entomologists' Monthly Magazine* addresses only one or two specific questions.

The detailed study of almost any animal has a fascination of its own. Hoverfly adults are suitable for studying many aspects of ecology, particularly feeding and mating behaviour. The larvae are good subjects for other aspects of ecology, such as detailed movement patterns, and the behavioural elements of feeding on aphids.

A common problem with biological studies is that they focus upon a particular stage of the life-history, and often fail to integrate this knowledge into the biology of the species as a whole. For example, many aphid-feeding hoverfly larvae appear to eat only a restricted number of aphid species, judging by where they are found in the field. In the laboratory, larvae will eat almost any aphid offered to them. Specificity in the field appears primarily because females lay eggs only in the colonies of certain aphid species, and is apparently not due to feeding preferences of the larvae. Thus patterns seen in the field can have complex causes.

Table 1. *Classification of the family Syrphidae, including only species dealt with in this book. Names largely follow Stubbs & Falk (1983)*

Subfamily, tribe, genus and species	Plate
Subfamily Syrphinae	
Tribe Syrphini	
Syrphus ribesii (L.)	1.2
Syrphus vitripennis Mg.	1.1
Epistrophe eligans (Harris)	1.7
Metasyrphus corollae (Fabr.)	1.10
Metasyrphus luniger (Mg.)	1.9
Scaeva pyrasti (L.)	1.8
Dasysyrphus albostriatus (Fall.)	1.3
Dasysyrphus tricinctus (Fall.)	1.4
Dasysyrphus venustus (Mg.)	1.5
Leucozona lucorum (L.)	2.2
Meliscaeva auricollis (Mg.)	2.4
Meliscaeva cinctella Zett.	2.3
Episyrphus balteatus (DeGeer)	2.1
Sphaerophoria scripta (L.)	1.6
Chrysotoxum cautum (Harris)	2.13
Tribe Bacchini	
Baccha obscuripennis Mg.	2.5
Melanostoma mellinum (L.)	2.7
Melanostoma scalare (Fabr.)	2.6
Platycheirus albimanus (Fabr.)	2.8
Platycheirus clypeatus (Mg.)	2.12
Platycheirus manicatus (Mg.)	2.10
Platycheirus peltatus (Mg.)	2.11
Platycheirus scutatus (Mg.)	2.9
Subfamily Eristalinae	
Tribe Cheilosiini	
Cheilosia paganus (Mg.)	–
Rhingia campestris Mg.	3.3
Ferdinandea cuprea (Scop.)	3.2
Tribe Chrysogasterini	
Chrysogaster hirtella Loew	3.10
Neoascia tenur (Harris)	3.8
Neoascia podagrica (Fabr.)	3.9
Tribe Volucellini	
Volucella bombylans (L.)	4.7
Volucella pellucens (L.)	4.8
Tribe Sericomyiini	
Sericomyia silentis (Harris)	4.6
Tribe Xylotini	
Syritta pipiens (L.)	3.5
Xylota segnis (L.)	3.4
Xylota sylvarum (L.)	3.6
Tribe Eumerini	
Merodon equestris (Fabr.)	3.7
Tribe Eristalini	
Helophilus pendulus (L.)	3.1
Eristalis arbustorum (L.)	4.3
Eristalis intricarius (L.)	4.4
Eristalis pertinax (Scop.)	4.2
Eristalis tenax (L.)	4.1
Myiatropa florea (L.)	4.5

2 Some old friends

2.1 Introduction

The purpose of this chapter is to introduce some very common hoverflies, and to recount their life-histories. This will lead into the generalised account of hoverfly biology in the next chapter. The lives of these common species are fairly well understood in their basic outlines, and they demonstrate just a few of the amazing adaptations of insects to their environment. Four species have been chosen to illustrate the diversity of the family. In general, closely related species have similar life-histories, but usually there are some interesting differences too. Some of the differences are described at the appropriate points. Much of the detail of the stories is unknown, and provides fertile ground for further study.

2.2 The Drone-fly, *Eristalis tenax*

'Out of the eater came forth meat, and out of the strong came forth sweetness' (Judges 14:14)

The answer to Samson's riddle was of course that honeybees had produced honey while living in the carcass of a lion. Baron Osten Sacken devoted not a few pages to arguing that, in many cultures before the rise of scientific classification, *Eristalis tenax* was confused with the honeybee (Osten Sacken, 1894). The English name, the Drone-fly, also applied to other *Eristalis* species, stems from their resemblance to honeybee drones (see plate 4.1). The larvae of Drone-flies, Osten Sacken claimed, lived in rotting carcasses, and therefore this hoverfly, and not the honeybee, formed the basis for the riddle. Whatever the truth of the matter, the larval habits of *E. tenax* from our point of view can only be described as disgusting! Humans produce many rather disgusting habitats, and Drone-flies have exploited this to the full. The larvae can be found in drains, 'foul ooze', in streams and ponds, rotting matter, sewage, farmyard manure, and other similar places.

The larvae are more commonly known as 'rat-tailed maggots' from their appearance (fig. 5). They have a very long 'tail' with three tube-like sections that telescope in and out; when fully extended the tail can be four or five times as long as the body. At the end of the tail are eight oiled feather-like hairs that keep water out of the tube when submerged, and break through the water surface when the larva is in its normal position: for the tail is, of course, a type of snorkel, allowing the animal to take in air from the surface while feeding on the bottom. In the foulest of foul

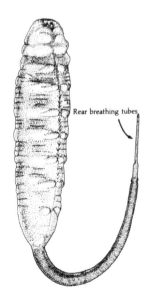

Fig. 5. The 'rat-tailed' larva of *Eristalis tenax*, the Drone-fly.

Rear breathing tubes

water one can often see great forests of these breathing tubes, waving slowly in the currents. The rather beautifully arranged mouthparts, pictured by Roberts (1970), are exquisitely adapted for feeding by filtering particles from the water. They consist of both coarse and fine filters. Water is sucked in and blown out again through the mouth by a muscular pump, and ejected water cleans the coarse filter to prevent clogging. Rat-tailed maggots also have a set of 'gills' in the rectum, that can be extended or retracted as required. Freshwater animals have a constant battle to keep water out of and salts in their body tissues, and these 'gills' actively take up salts from the water (see Schmidt-Nielsen, 1975, for an explanation of salt and water regulation).

The larvae of *Eristalis* could be used as indicators of pollution, since they only occur where they can filter large quantities of bacteria from the water. A project is now under way in Egypt to use them in just this manner.

Osten Sacken could find only one record of *E. tenax* larvae being found naturally in a rotting carcass, but it is easy to breed their relatives on submerged decaying animal matter. If this was the original food for larvae, then the range of man-made habitats in which they are now found is a testament to the superb adaptability of the species. There is one extraordinary and rather unlikely report in the literature of 'paedogenesis' by *E . tenax* larvae (Ibrahim & Gad, 1975). This is a process whereby a larva 'gives birth' to other larvae without any intervening adult stage, and is known from certain groups of animals, such as the mushroom fly, *Heteropeza*. Unknown from any other hoverfly, this might be a potent factor in the success of *E. tenax*.

The Drone-fly has been phenomenally successful. In about 1870 it spread via Asia to western North America, and then to the east as human communication became continuous across the continent. It was abundant right across North America by 1884. It reached New Zealand in 1888, and was common there the following year.

Mature larvae crawl out of the water and pupate just beneath the soil surface. After about two weeks the adults emerge from the pupa, usually in the morning. They remain motionless for a few hours while their wings expand and dry, and the skin toughens . They feed first on pollen, which contains the nutrients necessary to mature the reproductive system. Then the males tend to take mainly nectar from flowers to provide them with the energy they need for activity, whereas females take nectar and pollen in different ratios according to whether or not they are maturing a batch of eggs. The long 'tongue', or proboscis, allows them to feed from rather deep flowers that contain more nectar than shallower ones. Flowers of the Compositae (the daisy family) are commonly visited, particularly yellow chrysanthemums or asters; *E. tenax* is known to prefer yellow flowers (Ilse, 1949), and this can lead to the pollination of yellow-flowered crops (Kay, 1976). *(Eristalis* species are important pollinators of many flowers. Because they are large and hairy, pollen sticks to their bodies, and is thus transferred.) In late autumn Drone-flies become very common on ivy blossom. Occasionally they have been shown to transmit plant diseases, such as fireblight to apple trees.

Adults come in light or dark forms, light forms having much larger orange markings than dark forms. The frequency of light forms is higher in males. These differences are influenced mainly by a single gene, but environmental factors are also important since if larvae are kept at low temperatures, emerging adults are darker than normal (Heal, 1979). Honeybees show similar variations in colour pattern, and in general Drone-flies are good mimics of them.

Mimicry in *Eristalis* is certainly effective protection against some predators. Toads exposed to honeybees avoid trying to eat Drone-flies, whereas inexperienced ones eat them with relish (Brower & Brower, 1965). When you catch *E. tenax* in a net, it produces a very dangerous sounding buzz. In the nineteenth century it was believed that flaps of 'skin' in the breathing holes of the thorax produced this sound, and that other Drone-flies heard it by means of the 'plumes', feather-like structures near the halteres (fig. 2), that are unique to hoverflies. We now know that the sound is produced mainly by the wings, and that hoverflies can feel vibrations through their perch, using specialised organs in their feet. It is unlikely that airborne sound can be detected. However, we still have no idea what the 'plumes' are for.

Some fascinating experiments carried out on *Eristalis* have concerned mating behaviour. Males use different ways of courting and mating with females. In Canada, observations suggest that males remain all their lives in a certain defined area, within which they rest, shelter, groom, bask, feed and mate. In this 'home range' is a specific mating territory that is defended against all comers, including bees, wasps and other hoverflies (Wellington & Fitzpatrick, 1981). Several intruders were seen killed by territorial males. Such mating territoriality has rarely if ever been reported from Britain. Male *Eristalis* of several species, including *E. tenax*, search for mates near flowers; they court females by hovering a few centimetres above them while they feed, emitting a distinct and unusual flight tone. Males of other species, such as *E. pertinax* and *E. intricarius*, hover in large air spaces, occasionally darting away to chase passing insects. They are trying to catch females, but since most passing insects are not females of their own species, the chases are usually unsuccessful. They then return to hover in almost exactly the same place as before. The accuracy of the return is the same however long the chase was, or from whatever direction they return. Males thus have a sophisticated memory of their surroundings (see Collett & Land, 1975*a*).

When chasing, males of these species do not head towards their quarry and continually change direction as the position of the quarry changes, as in the case of *Syritta* (see Section 2.4). Instead, males actually calculate an interception course, and fly along it with a constant acceleration. If the object being chased suddenly reverses direction (as in experiments with corks on strings), males continue on the interception course without noticing that anything is amiss for about a tenth of a second. For this interception method to be successful in catching females, the males must 'assume' certain pieces of information. They 'assume' that they first see females at a distance of about a metre, and that females always travel at a constant speed (of about 8 metres per second). This information allows them to calculate an interception course from the rate at which the supposed female appears to be moving across the facets of the male's eyes. The course is only correct if he chases a female of his own species! For more details, consult Collett & Land (1978).

In late autumn, female *E. tenax* from the last generation of the year mate, and then find a place to overwinter, for instance in rock crevices, garden sheds, caves, houses or piles of dead leaves. Some may emigrate to southern Europe. A few males may also overwinter. The eggs of the female remain small and undeveloped, and the female's body is full of fat reserves. Sperm remain alive, nourished in some way by the female, and fertilise the eggs in spring. Thus very soon after emerging from their overwintering sites females lay batches of from 80 to 200 eggs, usually in a crevice near a source of food for the larvae, and the cycle begins again (Kendall & Stradling, 1972). Two or three generations are usually produced each year, and numbers may be augmented by immigration from the continent.

2.3 *Syrphus ribesii*

While walking through woodland you may have noticed a high-pitched whine coming from the nearby foliage. It is often difficult to tell from which direction the sound comes. In certain places the noise can be very loud - a veritable insect orchestra. In fact this 'singing' is produced by males of the species *Syrphus ribesii*, perched on leaves or twigs from 30 to 250 centimetres from the ground.

Fig. 6. Larva of *Syrphus ribesii*.

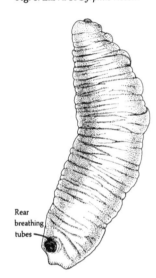

Rear
breathing
tubes

S. *ribesii* (pl. 1.2) is a widespread and very common species. The larvae are often the commonest predators to be found in colonies of many species of aphid, and are important elements in the natural control of these sometimes damaging pests. Being easy to rear given enough aphids, and having a rather limited range of easily observable types of behaviour, the larvae are very convenient subjects for research. The mature larva varies in colour from yellow to a delicate shade of pink on one part of the body, and is about 12 millimetres long and 3 millimetres wide (fig. 6). Those interested in larval structure should look at Rotheray's (1989a) book in this series.

What a rapacious predator the larva is! Its internal mouthparts are adapted for piercing aphids, and resemble a bird's beak. A newly hatched larva is tiny, and seems to find difficulty in piercing its prey. Once it has managed this, it can spend 8-12 hours sucking out the body contents. Older larvae are fearsomely efficient. They cast about in a semicircle, rearing the head up and then whipping it down again. Should the head contact an aphid, the mouthparts rapidly pierce the prey, and the larva often rears up again, lifting the aphid off its feet. The struggling aphid is then sucked dry, and the empty cuticle tossed aside. If you only watch larvae during the day, you might be forgiven for disbelieving this account of their efficiency. Normally sluggish and inactive during the daylight hours, at night they are frenetically active, feeding rapidly. If larvae are well fed they do not extract all the contents of each aphid, but discard them half-eaten. Feeding behaviour on aphids is described in more detail by Rotheray (1989a). There are many fascinating details still to be discovered about larval feeding behaviour. For example, Gerhard Gries filmed the behaviour of *Episyrphus balteatus* larvae in Germany, and discovered a remarkable fact. On encountering an aphid colony, the larva quickly moved over the colony, using its saliva secretion (that helps it move on the plant) to stick the aphids down. Having immobilized the aphids, the larva then settled down to feed !

The larvae have many enemies, particularly among parasitic wasps. A female wasp finds larvae by searching aphid colonies, detecting her prey using the antennae. The wasp usually climbs on the back of the larva, pierces it with her ovipositor (egg-tube), and lays an egg inside the body cavity. Some wasps lay just behind the mouthparts, taking advantage of larval defences to position the egg correctly. They bite the larval skin, causing the larva to raise its head; the wasp then smartly sticks in her ovipositor and lays (see Rotheray, 1979,1981a). A larva defends itself either by rearing up and spitting (see below), or by preventing the wasp from mounting on its back by spasmodically rolling over and over. In the laboratory the battle is almost always won by the wasp, but in nature this rolling behaviour would probably cause the larva to drop off the plant and escape.

Ants are often associated with aphid colonies, carefully tending and 'milking' the aphids for their honeydew (see Dixon, 1973). Contacts between larvae and ants are therefore likely to be common, and it is claimed that *Syrphus* can produce a sticky salivary 'glue' as a defence. When attacked, the larva arches its back to place its mouthparts on the ant's body, and spits out a drop of fluid. The ant immediately stops attacking, and tries to remove the entangling fluid (Eisner, 1971).

When the larva is fully grown it gets rid of the black contents of its stomach, and pupates on the ground or hidden in the vegetation. If it matures late in the season it does not pupate, but looks for a hibernation refuge, usually among fallen leaves. All further development virtually stops, and the larva enters a resting stage (diapause). In spring, development resumes, and pupation takes place. Some closely related species can overwinter in different stages of the life-cycle: for example, *Metasyrphus corollae* overwinters as a pupa, and *Episyrphus balteatus* as an adult. Another way of dealing with the rigours of winter is to migrate as an adult. A large proportion of the North European populations of *E. balteatus*, *M. corollae* and S. *ribesii* migrates in huge numbers to the Mediterranean, flying over the high passes of the Alps (Aubert and others, 1976; Gatter & Schmid 1990).

The pupal stage lasts only a few days, and the adult soon emerges, normally in the morning. The pristine fly climbs up the vegetation, and waits while its wings expand and harden. The striking black and yellow colours take a few hours to develop fully.

Adults feed initially only on pollen, which provides the nutrients needed for maturing their reproductive systems. Females continue to feed mainly or exclusively on pollen throughout the few days to weeks that they remain alive. Once capable of mating, males spend most of their time attempting to capture and mate with females, and therefore feed mainly on honeydew or nectar, both of which are readier sources of energy than pollen. Adult S. *ribesii* have such short mouthparts that they are unable to reach nectar in any but the most open of flowers. Thus when the hawthorn bushes are in bloom males will take nectar, but later in the season they are usually seen feeding on honeydew from leaves.

This brings us back to the 'singing' of the males. Looking closely, one can usually see that the wings are blurred during singing, indicating that they are vibrating rapidly: this is the source of the sound. It occurs when the huge flight muscles of the thorax are working but their coupling to the wings has been modified so that the muscles produce heat and the wings whirr, but the insect does not fly. Singing is in fact a method of raising the thoracic temperature, necessary because flight is impossible or very inefficient if the temperature of the flight muscles is too low. If you listen carefully, you can hear the pitch of the note rise as the thorax gets warmer.

Males congregate in swarms under trees, and chase passing insects either from a perch or from a hovering station. Perching is the usual technique when air temperature is low, and hovering is used when it is high. A male must be quick off the mark to get to a female before the others, hence the importance of keeping the flight muscles warm while perching. S. *ribesii* shows a remarkable ability to regulate its thoracic temperature for so small an insect, and this ability is probably associated with the very active and agile flight of these flies. Heinrich's (1979) book on bumblebees gives a vivid picture of the benefits of temperature regulation in insects.

While hovering, males will allow you to approach within a few centimetres if you move slowly and smoothly, and they will often perch on an outstretched finger. Because they will do this, but do not move to the edge of the air space to perch, their position in the air space must in some way be important to them. Perhaps a central position allows them to reach a passing female more quickly.

You will be very lucky to see an actual mating in this species, for it probably occurs in flight, and is (in captivity) extremely quick, lasting for less than 2 seconds. However, a recent observation shows that the flies also mate while resting on twigs. Females probably mate while their eggs are undeveloped, and mating may influence their final maturation. When the eggs are fully developed, females are attracted to aphid colonies, mainly by the smell of honeydew. They run about over the leaves with their mouthparts pressed to the leaf surface, tasting or feeding on the honeydew. Young females are very discriminating about where they lay, laying eggs only in or near aphid colonies; older females tend to lose this discrimination, and will lay in the absence of aphids. Larger colonies of aphids have a greater effect in stimulating females to lay. Eggs are laid singly, or occasionally in pairs.

2.4 *Syritta pipiens*

Perhaps the commonest of hoverflies whose larvae do not feed on aphids, and one of the most ubiquitous of all, *Syritta pipiens* (pl. 3.5) is an ideal subject for study. It is an engaging little fly, not easily disturbed by a human presence, and hence close observation of its behaviour is possible. *Syritta* is a most skilful flier, being able with equal ease to hover, or fly forwards or backwards. Males are most

Fig. 7. Larva of *Syritta pipiens*.

Rear breathing tubes

in evidence, displaying their acrobatic skills around virtually any clump of flowers.

The larvae are characteristically to be found in garden compost heaps, silage, manure, and other places where animal or vegetable matter is decomposing. They are of the 'short-tailed' type (see fig. 7), rather than the 'long-tailed' type like *Eristalis*. When fully grown the larva is about 10 millimetres long and almost 3 millimetres wide. The front end is formed into what is described as a 'false head' (whose function is probably to help maximise the amount of food that can be taken in), there are seven pairs of false legs along the body, and the 'tail' is about a millimetre long. Larvae have been found in December, emerging in May, and pupae have been found in October. It is therefore unclear whether *Syritta* overwinters as a mature larva or as a pupa.

The adults are small, and use their very fine proboscis to probe for nectar in small flowers such as forget-me-not (*Myosotis*) and some Compositae. Females take a good deal of pollen, but males take mostly nectar, in accordance with their highly energetic mating behaviour. All day long males cruise around flowers, searching for females. They chase other flying insects, and leap on resting ones, even bumblebees. This behaviour has been interpreted as territorial defence, but it may be that males approach every insect as a female *Syritta* until they find to the contrary. Females appear to play no role in the courtship, and the behaviour of males has been termed 'rape'.

As in *Eristalis*, the chasing behaviour of male *Syritta* has been analysed, showing some fascinating features. Males chase other insects by keeping about 10 centimetres behind them ('tracking') for many seconds until the chased insect lands, permitting an attempted 'rape'. Unlike *Eristalis* they track their quarry by continually altering the angle of their flight as the position of the chased insect changes, and they do not calculate an interception course. The tracking is done, as in other male hoverflies, using the large facets of the compound eyes that point forwards, those at the place where the eyes meet (fig. 15, p. 37): females lack these facets (fig. 16, p. 37). These large facets produce acute vision because of the physics of this type of eye, which affords a specialised area with greater visual acuity, analogous to the fovea of vertebrates. The greater accuracy probably allows males to keep the correct distance from their quarry by estimating the apparent size of the image on the retina. This is only possible because of the male's slow, stealthy pursuit, which depends upon not being seen by a female until it is too late. By contrast, *Eristalis*, *Syrphus ribesii* and other hoverflies rely on an ability to overtake females rapidly, pursuing and capturing them in flight. The speed of this manoeuvre means that a small fovea is impractical, and they appear not to possess a well-defined foveal region.

When a male jumps on a resting female he often does not fly straight towards her, but 'wiggles' rapidly in the last 50 centimetres of the approach. This in-flight oscillation may also serve to identify males to one another in head-on encounters. Both start oscillating in opposite directions, with the size of the oscillations increasing, until they break off and one flies away. This extraordinary behaviour may serve other functions that remain to be discovered. For more information see Collett & Land (1975*b*).

Needless to say, mating is over quickly during the 'rape'. *Syritta* is the only species in which a female has been seen to mate with more than one male in the wild although this is generally the rule for laboratory cultures of all species. When females are ready to lay their eggs they fly to a suitable site, crawl into the debris, and lay a batch of up to several hundred eggs. Males have been seen to hover over females while they do this, perhaps guarding them to prevent other males from attempting to mate.

2.5 *Volucella* species

It is perhaps of *Volucella* that Thomas Muffet wrote in 1658: *'it is swifter of wing than all the rest . . . sometimes it sits in one place for a great while together, as if it were unmovable, but as soon as you come near it, its out of your sight before you can say, What's this? and will not yeeld a jot to the Swallow (from whom it hath its name) for swiftnesse of flight'.*

The genus *Volucella* is in several ways the most interesting and important of all hoverflies. This is because their larvae show a range of key evolutionary innovations which led to the evolution of many different genera. We have five species in the UK: *bombylans, pellucens, inflata, inanis* and *zonaria*.

All *Volucella* species are powerful fliers, with recorded steady flight speeds of 3.5 metres per second, and much faster speeds over short distances. As far as is known, male *Volucella* (pl. 4.8) hover in exactly the same manner as *Eristalis pertinax* and *E. intricarius*. The mating behaviour of many species is not well described.

In common with species of *Microdon* (a rare genus not treated in this book), the larvae live in the nests of social insects. Unlike *Microdon* (whose larvae are 'guests' in ants' nests), *Volucella* larvae live in the subterranean nests of bumblebees and wasps. You might imagine that the mimicry of bumblebees evident in the hair colours of *V. bombylans* (pl. 4.7) is useful in deceiving the bees as female flies lay eggs in their nests. Apparently this is not the case. In the nineteenth century Jean-Henri Fabre found that no amount of resemblance to wasps would save a fly placed on the combs of a wasp nest (the usual host of *V. pellucens* and *V. zonaria):* flies were immediately attacked and killed. In fact, female *Volucella* have been seen to enter the nest at dusk, and to keep to the outer surfaces of the nest, never contacting the comb: they lay eggs there undisturbed by the nest's occupants. Even if stung, female *Volucella* are able to lay their eggs while dying. The fact that they can get into the nests, and remain there undisturbed, points to the possibility that they produce certain chemicals that mislead the defenders into believing that they belong to the nest.

We know the larvae of four of the five UK species - only *V. inflata* has eluded all efforts to find it. This is a shame, because it is probably the most critical species of all in enabling us to understand syrphid evolution. The larva of *V. inflata* is said to live in large sap flows, such as those resulting from the boring of goat-moth caterpillars in wood.

Unlike nearly all other genera, which have few larval differences between their constituent species, *Volucella* larvae fall into two types (and perhaps a third, when the larva of *V. inflata* is discovered!). *V. zonaria, bombylans* and *pellucens* are all of the same type: large, with numerous pointed projections. *V. inanis* is completely different, being flattened and relatively featureless.

On hatching, larvae of the *pellucens* type fall to the bottom of the nest, where a rich provender of dead wasps (grubs and adults) accumulates from the hygienic operations of workers. By watching nests provided with glass tops, Fabre saw some *Volucella* larvae wandering over the comb. These larvae actually entered the cells containing wasp grubs, and seemed to be eating the latter's excretory products. Such larvae were unmolested by adult wasps, whereas the larvae of other flies placed experimentally on the comb were attacked and killed. Perhaps *Volucella* larvae produce a chemical that 'appeases' the wasps: such a chemical is known to be produced by some beetles that live in ants' nests. We don't really know whether these larvae are scavengers, or true predators, or perhaps both.

In complete contrast, *V. inanis* larvae insinuate themselves inside the brood cells, between the wasp larva and the cell wall. Their very flattened form allows them to do this. They then fasten their mouthparts on to the wasp grub, and feed on it. Eventually the larva sucks out all the body contents of the grub, for this hoverfly larva is an ectoparasitoid.

Adult *V. bombylans* are often found in the same habitats and at the same time as bumblebees, and their resemblance to certain bumblebee species is striking. Across the geographical range of this species, from North America to Europe and Russia, there are several colour varieties, each of which mimics a different bumblebee species that occurs in the same area. In Britain we have two main forms that resemble closely two of our bumblebees. The protective advantages of these resemblances have not been demonstrated in *Volucella* (see Section 3.6.4).

In the genus *Volucella* we have an example of a species that has expanded its range within the last 50 years, colonising Britain in the process. *V. zonaria* is the largest and perhaps the most magnificent British hoverfly; it was a very rare straggler to these shores until recently. Then, a rash of captures indicated that this species might be breeding in the southern counties, a possibility subsequently confirmed by breeding records. *V. zonaria* became fairly common during 1940-60, its abundance varying with wasp abundance (Smith, 1974), and it was captured as far north as south-west Essex. Recently, it appears to have become rarer, for unknown reasons.

3 The biology of hoverflies

3.1 Introduction

Hoverflies are an interesting group to study, not only because they are large and colourful, but also because they have such varied life-histories. As with nearly all insect groups, comparatively little is known of their biology and ecology. The larvae of a large percentage of species are unknown, even in Britain and other European countries: the life-history of most species found outside Europe has yet to be described.

The book by Imms (1971) gives a popular account of the diversity of insects and their life-histories and introduces the Diptera (flies), describing how they develop and function. Books by Chinery (1976) and Colyer & Hammond (1968) may also prove useful.

All flies pass through larval and pupal stages as they develop from egg to adult. During the pupal stage the organs of the larva are almost completely broken down and the adult body assembled. The time needed for development from egg to adult varies greatly between species, taking less than 2 weeks in some hoverflies and possibly up to 5 years in others. With such a large divergence in types of hoverfly life-history, it is difficult to generalise about development. In addition, we know only about the commonest species. A great deal of valuable knowledge can be obtained merely by rearing species to adulthood and describing the stages of development. This chapter describes the biology of hoverflies, and tries to show where gaps in our knowledge can be filled by your studies. The methods that are mentioned as being of possible use are described in chapter 5.

3.2 The evolution of hoverflies

The pattern of evolution in syrphids is now reasonably well known from detailed studies of larval structure (Rotheray & Gilbert 1989, 1993). The most 'primitive' type is *Eumerus* (not treated in detail in this book). From this type, those with plant-feeding larval forms such as *Merodon* and *Cheilosia* arose, close to the root of the evolutionary tree. Also fairly primitive are the predatory types: *Volucella, Microdon,* and the aphid predators. The main trunk of the tree contains forms such as *Rhingia, Ferdinandea, Chrysogaster, Neoascia* and *Syritta.* There is then a fork in the trunk, with one branch leading to various fairly rare forms such as *Criorhina* and *Callicera,* and another leading to *Sericomyia, Eristalis, Helophilus* and *Myiatropa.*

3.3 The egg

Eggs can be found singly, or in batches of up to several hundred. Batch sizes are normally characteristic of the species, and are related to the larval feeding habit. Thus, in general, species with aphid-feeding larvae lay eggs singly or in groups of two or three, whereas species such as *E. tenax* that filter food from water can lay up to 200 per batch.

All known eggs have a similar shape - an elongated ovoid with one end narrower than the other. They vary in size. For instance, *Volucella* species lay very large eggs, more than 2 millimetres in length, whereas the small *Syritta* lays eggs barely 0.5 millimetre long. Generally, larger species lay larger eggs, but there are exceptions to this. Certain related species (*Syritta* and *Xylota*, for example) lay eggs that are much smaller than might be expected for flies of their size. These differences are ecologically important, because egg size relative to the size of the female characterises the investment a female makes in her offspring in terms of the nutrients she uses to fashion each egg. Dissecting mature females, that is those with a batch of eggs ready to lay, and counting and measuring the eggs (see Section 5.4), will build up a picture of these relationships. Dissections done regularly throughout the season will determine whether eggs are produced only at particular times of the year, and whether egg development or the act of mating is associated with any other feature of individual specimens (such as the presence of pollen in the gut or crop).

Fig. 8. Examples of the intricate surface patterns of hoverfly eggs.

Epistrophe eligans

Syrphus ribesii

The simplest way to collect eggs is to catch mature females and allow them to lay in tubes that have been provided with suitable larval food. Very often the mass of developed eggs can be seen through the thin sides of the abdomen, allowing the assessment of maturity in the field. With this method, there is no problem about identification of larvae, since the adult can be used. If access to a good compound microscope is possible, have a look at the beautiful and delicate sculpturing on the egg surface (fig. 8). Chandler (1968*a*) has produced a key to the eggs of some of the common hoverflies that relies mainly on differences in surface patterning. Chandler also shows how to distinguish hoverfly eggs from those of other insects, and with practice this can be done in the field without a microscope. It is therefore possible to identify eggs found in the wild. Recording the positions of eggs relative to various factors (such as aphid colonies or leaf midribs) might produce some interesting results. Some work of this nature has been carried out. Chandler (1968*b*) and Smith (1976) show that some species (*Melanostoma*, most *Platycheirus*) lay on leaves irrespective of whether aphids are present or absent, whereas most species (whose larvae can *only* eat aphids: see Section 3.4.1) lay next to or in aphid colonies.

Fig. 9. Larva of *Epistrophe eligans*.

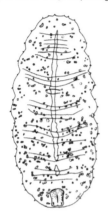

The egg stage is very brief in common species, typically lasting less than 5 days, but this varies with conditions of temperature and humidity. High temperatures and humidities accelerate development and hatching.

3.4 The larval stage

Hoverfly larvae can be divided into two broad groups: the carnivorous ones and the rest. As described in the next section, the 'rest' can be further divided into three groups. The carnivorous species are unique among fly larvae in that they are coloured; most other fly larvae are white or dirty yellow (like the maggots used for fishing). Observe one of these carnivorous larvae under a binocular microscope. The beautiful colours are produced by blood pigments or by fat deposits under the skin, and the resulting colour pattern often hides the larva - there is even one species whose pattern mimics a bird-dropping (Rotheray, 1989b). Non-carnivorous hoverfly larvae are usually white. Some larval forms have been described in chapter 2 (figs 5-7), and others are depicted in figs 9-12.

As noted for S. *ribesii*, hoverfly larvae make very good subjects for experimentation, being easy to obtain and rear. They can, for example, easily be kept in petri dishes. Little research has been done on non-carnivorous larvae, and although more is known of the carnivorous ones, there is still plenty of scope. In particular we need to be able to compare different species; at present we understand a few aspects of the biology of a mere dozen species or less.

Fig. 10. Larva of *Chrysogaster hirtella*.

Fig. 11. Larva of *Merodon equestris*.

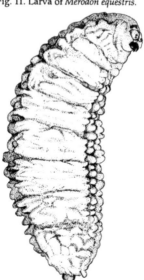

Fig. 12. Larva of *Rhingia campestris*.

3.4.1 *Feeding habits*

The larvae display an astonishing range of feeding habits, broader than those of most other insect groups. They can be considered in four categories that grade into one another: species that feed on plant tissues and plant products (such as sap), species that scavenge on or filter decaying matter (usually in water), species that live in the nests of social insects (bees, wasps and ants) and carnivores, whose normal food is aphids.

Some species from Britain and other parts of the world have rather exotic feeding habits: for example, some feed on maize pollen, or in pitcher plants, in decaying cacti, in mushrooms, in truffles, or on aphids in tree galls. Among carnivorous types alone there is a large range of prey types recorded, from thrips to blowflies (see Ghorpade, 1981, for a review). The larval feeding habits of the species covered in this book are summarised in table 2. Note that *Eristalis* and *Helophilus* larvae filter their food from stagnant and filthy water in drains and ponds. In countries where people have to drink unpurified water, swallowing eggs or small larvae has caused alarming symptoms of nausea, giddiness and even more serious problems (Zumpt, 1964).

From man's point of view the most important species are those whose larvae feed on aphids; from table 2 you can see that all these species belong to the subfamily Syrphinae (check against table 1).

We used to believe that some of these (*Melanostoma* species and some *Platycheirus* species) could, if no aphids were available, feed on rotting plant material. This is almost certainly incorrect: they cannot survive on this material. But they may well be generalist predators in the leaf litter, rather than only taking aphid prey. There are large numbers of other types of animals available in leaf litter - perhaps they feed there. It remains true that these species are among the commonest of all hoverflies as adults, and yet they are rarely found as larvae - where are they?

Many carnivorous species will accept a range of different aphids as food, but some aphids are definitely avoided, and at least one is toxic (Růžička, 1975). Catholic tastes in aphids are found in nearly all the common hoverflies (*Syrphus ribesii, Episyrphus balteatus, Metasyrphus corollae, Scaeva pyrastri,* for example). Rarer hoverflies with aphid-feeding larvae are often found on only a restricted range of aphids, or even on only one type. Larvae of different species may also specialise on a particular part of the plant, on which they are most efficient at capturing their prey. The question of why certain species are generalists while others are specialists is the subject of many ecological theories (see Ricklefs, 1991). Rotheray (1989a) discusses these problems in the context of aphid predators.

Table 2. *Common hoverfly larvae: feeding habits, where they are found, references to their descriptions*

Species	Larval food	Larval habitat	Description
Syrphus ribesii	Aphids	Trees, shrubs, herbs	Bhatia (1939)
S. vitripennis	Aphids	Herbs	Dixon (1960)
Epistrophe eligans	Aphids	Trees, shrubs	Dixon (1960)
Metasyrphus corollae	Aphids	Trees, shrubs, herbs	Goeldlin (1974)
M. luniger	Aphids	Herbs	Bhatia (1939)
Scaeva pyrastri	Aphids	Trees, shrubs, herbs	Bhatia (1939)
Dasysyrphus albostriatus	Aphids	Trees	Goeldlin (1974)
D. tricinctus	Aphids	Trees	Rotheray (1989*b*)
D. venustus	Aphids	Trees	Rotheray (1989*b*)
Leucozona lucorum	Aphids	Trees, shrubs, herbs	Dixon (1960)
Meliscaeva auricollis	Aphids	Shrubs	Dixon (1960)
M. cinctella	Aphids	Trees	Dixon (1960)
Episyrphus balteatus	Aphids	Trees, shrubs, herbs	Bhatia (1939)
Sphaerophoria scripta	Aphids	Herbs	Bhatia (1939)
Chrysotoxum cautum	Aphids	Probably underground (root aphids)	Rotheray & Gilbert (1989)
Baccha obscuripennis	Aphids	Herbs	Dixon (1960)
Melanostoma mellinum	Aphids ?	Herbs	Heiss (1938)
M.scalare	Aphids ?	Herbs	Dixon (1960)
Platycheirus clypeatus	Aphids ?	Herbs	Dixon (1960)
P.albimanus	Aphids	Herbs	Dixon (1960)
P.manicatus	Aphids ?	Herbs	Dixon (1960)
P.peltatus	Aphids ?	Herbs	Goeldlin (1974)
P. scutatus	Aphids	Trees, shrubs, herbs	Bhatia (1939)
Cheilosia paganus	Plants	Cow parsley (*Anthriscus sylvestris*)	Stubbs (1980)
Rhingia campestris	Dung	Cowpats	Coe (1942)
Ferdinandea cuprea	Sap	Sap flows of trees	Hartley (1961)
Chrysogaster hirtella	Filtered from water	Attached to plants in freshwater	Varley (1937)
Neoascia tenur	Decaying matter	Decaying *Typha*	Hartley (unpubl.)
N. podagrica	Decaying matter	Manure, silage	Hartley (1961)
Volucella bombylans	Dead grubs, faeces, etc.	Wasp and bumblebee nests	Dixon (1960)
V. pellucens	Dead grubs, faeces, etc.	Wasp and bumblebee nests	Hartley (1961)
Sericomyia silentis	Filtered in peaty pools	Bogs, marshes (*S.lappona*)	Hartley (1961)
Syritta pipiens	Decaying matter	Manure, compost, silage	Hartley (1961)
Xylota segnis	Decaying wood or leaves	Silage, rotting wood	Hartley(1961)
X. sylvarum	Decaying wood or leaves	Wet sawdust, under rotting bark, etc.	Hartley (1961)
Merodon equestris	Bulbs	Daffodils, bulbs, etc.	Hartley (1961)
Helophilus pendulus	Filtered from water	Drains, wet manure, cesspits	Hartley (1961)
Eristalis arbustorum	Filtered from water	Polluted water, drains	Hartley (1961)
E. intricarius	Filtered from water	Drains, stagnant water	Hartley (1961)
E. pertinax	Filtered from water	Drains, wet manure	Hartley (1961)
E. tenax	Filtered from water	Drains, sewage, polluted water	Buckton (1895)
Myiatropa florea	Filtered from water	Rotholes of trees, especially beech	Roberts (1970)

Each larva can eat up to 1200 aphids during the course of development. The actual number varies because aphids vary in size both between species and during their development. Some, such as *Syrphus ribesii*, are in general more efficient predators than other species such as *Melanostoma*, attacking aphids at a faster rate, and eating, moving and searching more rapidly. However, *S.ribesii* does appear to require more aphids to complete growth.

Aphid colonies are frequently obliterated by feeding larvae, resulting in periods of starvation for larvae. This can lead to an unusually small adult size and a consequent reduction in the number of eggs laid by a female (Cornelius & Barlow, 1980). Starved larvae 'attempt' to compensate for their nutritional deficiencies by searching harder, capturing prey more rapidly, and sucking 50% more of the contents of each aphid before discarding it (Leir & Barlow, 1982). A series of questions about the behaviour of aphid-feeding larvae has been listed in table 3. These are meant only to suggest possible experiments, and are not exhaustive. The experiments are important because after more than 50 years of study, entomologists do not agree about how effective hoverfly larvae are at controlling aphid pest populations. Attempts to introduce European species into Canada and Hawaii (for aphid control) and into the United States (plant-feeding species to control thistles) have all failed.

Table 3. *A behavioural inventory for aphid-feeding hoverfly larvae: some questions*

Prey
What is the prey?
What range of prey types does the individual/species use
Do larvae reject any prey?
What aphid instars are eaten by each larval instar? With what frequency?
How do larvae attack, subdue and eat their prey?
At what rate do larvae attack? Does this alter systematically when you increase prey density?
Are larvae feeding continuously, or do they have resting periods? Where do they rest?
How quickly do larvae eat? Do they suck prey out completely, or do they leave part uneaten?
Do aphids escape after contact with a feeding larva? How?
Are some aphids better at escaping than others?

How many aphids are needed to complete development?
Howdoes this vary according to aphid size?
Movement
How quickly do larvae move?
Is speed affected by what they are moving on?
Do larvae concentrate on searching a particular part of the plant? Is this where the aphids are?
Do they move between aphid colonies? Is this a response to starvation?
Interactions
Do ants tolerate the presence of larvae? Can larvae defend themselves?
Do you find more than one species of hoverfly in an aphid colony? Are larvae cannibalistic?
How are larvae killed?

We do not know enough about hoverfly biology to be certain why these introductions were unsuccessful.

A more detailed account of the biology of aphid-feeding larvae, together with further suggestions about experiments, can be found in another book in this series (Rotheray, 1989*a*).

Species with larvae that do not feed on aphids are varied in their larval biology. The larvae can be harder to find, though a search of suitable sites, selected by looking at table 2, should prove successful at the right time of year. Because they are more difficult to find, much less is known about these larvae, giving more scope for further research. There is one exception to this generality. The larvae of the bulb flies (Eumerini), which include *Merodon equestris*, are pests of commercial bulb plantings such as daffodils and onions, and because they have some economic importance more work has been carried out on their biology (see Doucette and others, 1942).

Although such larvae can be hard to find, sometimes they are much easier to find than the adults. We have had a rather dramatic case of this with the hoverfly *Callicera rufa*. This beautiful fly is very rare indeed in collections; in fact, the same is true of all the species of *Callicera*. The larvae live in rot-holes of Scots pine, and are only present in old stands. A survey of Scots pine all over Scotland revealed dozens of new sites where larvae were found, but no adults had ever been taken (Rotheray & MacGowan 1990).

3.4.2 *Length of larval life*

As do all insects, hoverfly larvae have a rather inextensible skin, the cuticle, that must be shed to allow growth. Thus all hoverflies pass through three stages, or instars, before pupating, shedding the skin between each instar and the next. The shed cuticle is thin and transparent, and is therefore difficult to see unless the larva is kept in a small, transparent container such as a petri dish.

The length of larval life varies enormously between species, partly because some but not all overwinter as third-instar larvae. Longevity is also bound up with the timing and length of the seasonal appearance of adults. Differences depend mainly upon the occurrence (or absence) of diapause, a resting period during which little or no growth takes place. Thus some species (*E. balteatus, M. corollae, S. ribesii*) are said to fit as many generations into the year as weather conditions permit, overwintering as adults or as non-diapause larvae. Such species can spend as little as 10 days as larvae, depending on temperature, humidity, food supply, day length, and other such factors (the influence of all these factors is not well studied). In spite of the very short larval stage of these species, there is little evidence that more than three generations are actually produced during any one year, and *E. balteatus* sometimes has only one.

In contrast to the above species, others (such as *Epistrophe eligans, Merodon equestris*) regularly have only one generation each year, adults usually appearing in the spring. The third-instar larvae enter diapause at some point during the year, overwinter as larvae, and pupate in the spring. Larval life is therefore very long, lasting almost a year, because development is more or less arrested during diapause and the low temperatures of winter. Overwintering larvae can often be found in the leaf litter at the foot of trees and shrubs.

Some species are known to have two generations per year, with both a summer and a winter diapause in larvae of different generations (for example, *Dasysyrphus albostriatus*).

Occasionally, reared larvae have lived for more than a year before pupation. For example, a small percentage of *Merodon equestris* larvae take 2 years instead of one to complete development; we can regard this as an 'insurance policy' against the possibility of poor weather destroying the current year's brood. *Callicera rufa* was kept as a larva for 5 years (Coe, 1941), but it is possible that the rearing conditions prevented this larva from pupating, since the normal developmental period for this species appears to be two years.

3.4.3 *Parasitism*

Parasites are rarely reported from species whose larvae do not feed on aphids. This section therefore refers only to aphid-feeding species.

If larvae are collected from the field, many, or even most, will be found to have been parasitised. Many emerging parasites are difficult to identify, but others leave distinctive holes in the pupal case, or affect the growth of the larva in a recognisable way. Nearly all will be wasps, and many will be of the group Diplazontinae, for which there is a key (Fitton & Rotheray, 1982). The Diplazontinae contains many species that specialise in parasitising hoverfly larvae.

Female parasites lay their eggs in hoverfly eggs or larvae, and their young feed on the tissues and organs of the host. Often the wasp lays only one egg per host. Occasionally two females lay in the same host, and the wasp larvae compete for possession. The transparent skin of most hoverfly larvae allows one to see the living parasite inside, and therefore to experiment with its behaviour (see Rotheray, 1989*a*).

The adult parasite usually emerges from the pupal stage of the host, cutting its way out through the thick cuticle. A particularly interesting example of a parasite that can be identified from the cut that it makes is described by Rotheray (1981b).

Descriptions of exactly how parasites lay eggs in larvae, particularly in the wild, are not common in the literature. The larvae can defend themselves actively by rolling over, or by producing a sticky or perhaps poisonous saliva, but little is known of the effectiveness of these mechanisms in the field (Rotheray, 1981a; Eisner, 1971). By sitting quietly by an aphid colony in which hoverfly larvae are feeding, you may record some fascinating observations that invite further research (see Rotheray, 1989a).

3.4.4 Identifying the larvae

Hoverfly larvae often have colours that are useful aids to their identification (Rotheray, 1989b). The best way of observing larvae of aphid-feeding species closely is to take them off the plant with a moistened paintbrush, put them on a microscope slide, and examine them under a binocular microscope. Permanent collections of larvae can be made by killing them in boiling water, and preserving them in 70% alcohol, but there is no known method for retaining the colours.

Some hoverfly larvae have been illustrated in figs 5-7 and 9-12. Identification is not difficult for many species, especially aphid-feeding ones. The most important character is the form of the breathing tube at the rear end, which has a distinctive shape and structure in most species. Rotheray (1989a, 1993) has written and illustrated manuals for identifying many of the common species. Keys to most genera can be found in Rotheray & Gilbert (1989, 1993), and these can be supplemented by keys to species in the papers by Dixon (1960), Hartley (1961) and Dolezil (1970).

3.5 The puparium

When fully mature, the larva clears its digestive system by expelling a black oily liquid, and then it seeks a suitable site for pupation. Many species pupate just below the soil surface; others choose the lower surfaces of leaves, leaf litter, or other sheltered places. The larval skin of the third instar is not shed, but hardens instead to enclose the pupa. Pupation proper thus takes place within the last larval skin. Because what is seen is not the real pupa, strictly speaking this stage is called the 'puparium', to distinguish it from the true pupa of insects that shed the final larval skin.

Metasyrphus corollae and *M. luniger* pupate in late autumn and spend the winter as pupae, but apart from

Fig. 13. The pupa of *Episyrphus balteatus*.

these, most hoverflies have rather short lives (about 10 days) as pupae.

A pair of respiratory 'horns' projects from the pupa of many species. The breathing tube of the larva is also present in the pupa, affording characters that permit identification. The pupae of many aphid-feeding species resemble fig. 13. Scott (1939) has devised a workable key to such pupae, but unless rates of parasitism are very high, it is probably better to wait for the adults!

3.6 The adult stage

Collectors have often noticed that male hoverflies are usually caught earlier in the season than females in many species, and this earlier emergence is confirmed by breeding experiments. Males are generally smaller than females, and complete development slightly faster. This allows males to feed and mature their reproductive systems before females appear, ensuring that potential mates are not missed (see Thornhill & Alcock, 1983). Emergence takes place early in the morning, allowing time for the soft cuticle to harden and the wings to unfold, expand and dry before the hazards of the day begin in earnest. The abdomen is at first collapsed, and the fly expands it to its proper shape by swallowing air. The gaudy colours only develop gradually over several hours, and adults caught at this stage can be very difficult to identify. During this time, adults cannot fly, and are vulnerable.

In some species, particularly *Eristalis tenax* but also *Episyrphus balteatus* and others, a proportion of females overwinter in the adult stage. Thus adults that appear during the first warm days of spring and feed from early-flowering shrubs and herbs are very often not recent emergences but old females. The tatty condition of the wings and body hair will identify such individuals.

In summer and autumn there are large movements of hoverflies that are migrating from and to the continent. These are usually species with aphid-feeding larvae, but also include large numbers of *Eristalis tenax*. Most of our British-born *Episyrphus balteatus* appear to leave these shores at the onset of cooler weather in autumn, with only a small proportion remaining behind to overwinter as adults. Migrating swarms can be very large; when they fly over the Alps or Pyrenees they are concentrated by the 'bottle-neck' effect of the passes, and staggeringly large numbers are recorded (Owen, 1956; Aubert and others, 1976; Gatter & Schmid, 1990).

Hoverfly adults have large eyes, not because they are nocturnal (an explanation for the large eyes of many animals) but because they are fast fliers that need acute vision. Unlike that of other flying insects, the head of a

Fig. 14. The flight-path of the tip of
the wing; the short lines give an
indication of the angle of the wing
during the movement.
(Reproduced with permission
from Ellington (1984). *Phil. Trans.
Roy. Soc. B* **305**, 57.)

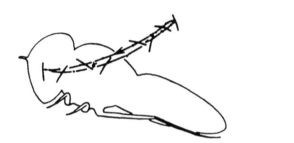

hoverfly remains absolutely still relative to the thorax while
the animal is flying. The superb acrobatic skill of these flies
is reminiscent of hummingbirds. In hovering, the wing
moves in a flat and oblique figure-of-eight, bending and
twisting during the stroke (fig. 14), and thus providing
plenty of lift. In cross-section the wing is not flat but
corrugated, and this improves aerodynamic performance.
You can see the corrugations by catching the light on the
wing surface. Under the microscope tiny hairs become
visible; these cover the wings of many species (but not
Eristalini), and probably alter air flow over the wing,
improving its performance. The thoracic muscles
responsible for the wingbeat (up to 250 beats per second)
comprise about 15% of the total body weight.

3.6.1 *Feeding*

Females emerge with completely undeveloped eggs, and
must feed to mature them. It is often impossible even to find
the ovaries of such females, let alone the eggs in them.
Males too probably emerge with an immature reproductive
system, and also need food to mature various parts of it
(principally the accessory glands). However, some mature
sperm may be present on emergence in some species, since
male *Metasyrphus corollae* are able to mate immediately they
emerge.
        For maturation feeding, the proteins and amino
acids of pollen are essential. Nectar, the other principal food,
contains only trace amounts of these substances, and

therefore cannot by itself be adequate to mature the
reproductive system. Once mature, male hoverflies probably
need rather little in the way of proteins and amino acids,
while the females still require large amounts for maturing
successive batches of eggs.

Pollen contains on average about 25% by weight of
energy-rich carbohydrates, and various lipids too, also rich
in energy. However, these sources are more difficult to
metabolise than simple sugars, and thus the energy in
pollen can probably be made available only slowly. Nectar is
an almost pure solution of sugars, and is thus a much better
fuel. For large species that use energy very quickly during
flight, nectar is a more important source of energy than
pollen.

Some small to medium-sized genera (such as
Melanostoma, Episyrphus) seem to feed almost exclusively on
pollen, obtaining from it all their energy and nutrients.
Other genera (such as *Eristalis, Metasyrphus*) appear to take
pollen for protein and nectar for energy. The mouthparts of
different species are adapted to these dietary differences.
Short, broad mouthparts are specialised for pollen feeding,
whereas long, thin mouthparts can reach the large amounts
of nectar in flowers with long corollae (Gilbert, 1981). The
complex structure of hoverfly mouthparts can easily be seen
with a microscope, and the interested reader should refer to
Gilbert (1981) for more details.

Many people have recorded the flowers that
hoverflies visit, but few have experimented upon the
reasons for their choice. Do hoverflies choose to visit certain
flowers and not others, or do they simply visit them in the
same proportion as they occur in the habitat? Hoverflies
may not visit some species because the flower is too rare,
blooms at the wrong time of year, is too small, has too little
pollen, or for many other reasons. We need to understand
how hoverflies themselves see flowers, and whether they
can distinguish between floral types. For instance, some
people have stated that most hoverflies visit mainly yellow
and white flowers, but within this category do not
distinguish between flower species, merely visiting them in
the same proportion as they occur in the habitat. Hoverflies
probably do select flowers predominantly by colour; scent
and other factors seem to play only a minor role. Most
hoverflies are recorded visiting composite, umbelliferous
and rosaceous flowers, selected mainly by colour, that is,
yellow or white flowers. A preference for yellow is known
in *Eristalis tenax* (Ilse, 1949; Kay, 1976).

Some hoverflies do, however, specialise on certain
plants. One group, consisting of the two *Melanostoma* species
and *Platycheirus clypeatus*, visits and takes pollen from wind-
pollinated flowers such as grasses and plantains. From a
series of elegant experiments these species are now known

to pollinate plantains, which are therefore not strictly wind-pollinated (Stelleman, 1978). Plantain pollen may even be adapted to being carried by insects in areas of low wind speed, since it is stickier in these areas. Other species also specialise on wind-pollinated tree pollen.

Rhingia campestris is another species that is rather choosy in the flowers it visits. It has by far the longest mouthparts of any British hoverfly (more than 10 millimetres), and it uses these to probe the deep, tubular corollae of blue or purple flowers such as Ground ivy or Bugle.

A final group of specialists is the *Xylota* species. They appear rarely to visit flowers at all, but instead collect pollen and other food items from leaf surfaces. The odd feeding habits of *Xylota* species may be connected with their remarkable mimicry of certain solitary wasps that move jerkily across leaves in a very similar manner. All *Xylota* feed in this way, even species from the USA, whereas no members of closely related genera behave like this.

We can take advantage of the feeding behaviour of hoverflies. Recent experiments have shown that, by planting flowers along the margins of crop fields, hoverflies are attracted in and lay more eggs on aphid colonies in the crop. This is an exciting area of biological control via habitat management.

Very little is known of the pollinating abilities of hoverflies. Several authors state that these flies are ineffective pollinators, basing their claim mainly on the fact that each species can be seen feeding from many different types of flower. Obviously this is not a realistic assessment. By following individuals and noting their visits, one can readily see that they do tend to visit flowers of one type in succession, and are therefore potential pollinators of many flowers, particularly composites and umbellifers. They are thus important economically, since they pollinate plants such as onions, carrots and fruit trees. In Japan, experiments have been carried out to assess the usefulness of *Eristalis* as a management tool in pollinating apple trees. *Eristalis* species are probably particularly effective pollinators because they have specialised body hairs that retain pollen, allowing it to be combed off and eaten during grooming at rest or during flight (Holloway, 1976). Many other plants are known to be pollinated mainly by hoverflies, including Bugle, Violet, several orchids, and even a palm! An objective index of the efficiency of the various visitors to insect-pollinated flowers would be the ability of the flower to produce seed after a single visit (Spears, 1983).

Table 4 lists a series of possible questions about feeding behaviour and pollination that might be tackled. These are not exhaustive. Proctor & Yeo's (1973) book gives an excellent background to pollination biology.

3.6.2 Mating behaviour

Mating behaviour serves to bring the sexes together, and to enable individuals to identify each other so that mating occurs between male and female of the same species. More complex functions of mating concern the *choice* of mate. Basically the idea is that females should be choosier than males because they make a greater investment in the offspring (it takes more nutrients to make eggs than sperm). Females should thus take care to mate with the 'best' available mate. Thornhill & Alcock's (1983) book provides an excellent account of the theory and reality of patterns of insect mating behaviour.

The mating behaviour of most hoverflies remains undescribed. It is common to see males of some species

Table 4. *Studies of feeding behaviour in hoverflies: some questions*

Feeding behaviour
What flowers are visited, and in what proportions?
  Do these proportions differ from the relative abundances of the flowers?
Do hoverflies differ in patterns of visits? Are these patterns related to characteristics of the flowers, such as corolla depth or colour?
Do hoverflies share the available flowers so as to minimise overlap in visiting patterns?
What foods are taken from each flower? Do species differ in the types of food they take?
Do individual flies visit only one flower species within one feeding sequence? How long does the sequence last?
At what rate are flowers visited? How long does it take to empty a flower of nectar or pollen?
Are individual flies often disturbed while feeding?
  What sort of disturbances occur? Do the flies return to the same flower?
Is there any evidence for patterns of dominance between hoverflies at flowers? If so, is this related to body size?
Do predators such as wasps take feeding hoverflies?
  How risky is feeding at different flowers?
At what time of day are flowers visited? Try marking individuals to see whether they visit the same groups of flowers on successive days.
Are visits timed to when pollen is liberated from anthers?

Pollination
Can you see contact between the stigma and the body of the fly?
  Is pollen transferred? (You could use coloured dusts as false pollen.)
Is the flower self-fertile or self-infertile, and hence what is the importance of insects as pollinators?
Are bumblebees more important in pollinating certain flowers? Honeybees?
With what frequency are individual flowers visited, and what visits them? What pollinates them?

---

hovering under trees in a shaft of light, the sun flashing from their bronze-coloured bodies. Females rarely, if ever, hover like this; energy from their food is destined for egg production. The hovering strategy is not followed by many species. In fact, three basic strategies seem to be used by male hoverflies. Many merely sit on twigs or leaves that afford a good view of the surrounding air space, and then chase any suitable-sized insect that flies by. Others hover in one position, singly or in small or large groups, rapidly darting away from the chosen spot to chase passing insects, returning when unsuccessful. The third method is perhaps the most common, and involves patrolling areas where females are likely to occur, such as feeding or egg-laying sites.

Males appear to be driven to find and mate with as many females as possible, and thus are probably less choosy than females. This behaviour causes them to test to the limits their tolerance of environmental factors. For example, on cold mornings in May one can often hear male *Syrphus ribesii* 'singing', as described in Section 2.3. They are attempting to raise their body temperature to a level sufficient for them to be able to fly after a female, should one chance by. Early in the morning, when it is cold, it is energetically costly for males to remain warm enough to be active: they test the limits of their ability to fly in cold weather.

Females are usually courted while feeding or resting, and do not normally appear to seek out males, except in species where males hover in swarms, such as S. *ribesii*. Females of this species probably visit the swarm to be mated, as in other swarming flies (such as mosquitoes).

Males that have been caught while searching for mates and then dissected will show a rather uniform anatomy. Usually there appears to be little but air in the abdomen, with the crop and genitalia residing in the tip. The crop and gut contain a viscous, concentrated sugar solution, with few or no pollen grains. Perhaps this enables them to carry as much energy as possible while remaining light enough for the energy not to be used up too quickly.

Once a male has found a female he must court her. Courtship behaviour has been described for only a few hoverflies. In at least one species (*Eristalis arbustorum*) the male hovers over a feeding female, his wings beating at an unusual rate so that the flight tone is different from normal. For many species, mating itself takes a few minutes at most, often only a few seconds or less while in flight. For others, mating may last several to many hours, and in summer it is common to see mated pairs of *Metasyrphus corollae* and *Sphaerophoria scripta* flying about and resting on leaves, the male riding on the back of the female.

Special tube-like glands in the female probably nourish the sperm while they are stored in her body. These glands are small in most species, but large in some *Eristalis* where the female mates in autumn and overwinters before laying. Sperm are stored in a group of three dark spherical structures called spermathecae (fig. 67, p. 58). In the Syrphinae there are two on one side of the abdomen, and one on the other. In all other hoverflies, all three are grouped together in the centre of the abdomen. They are connected by tiny ducts to the tube down which the eggs pass before they are laid. These ducts are little wider than the sperm. Sperm probably get into the spermathecae by suction. If you should dissect a female caught while mating, the spermathecae are usually very large and inflated, and their expansion probably sucks sperm up the ducts.

Table 5 lists a series of possible questions on mating behaviour. Current interest centres on whether female choice occurs, and if it does, what the basis is for the choice. The book by Krebs & Davies (1987) may also be useful, in addition to Thornhill & Alcock (1983).

Table 5. *The mating behaviour of hoverflies: some questions*

Mate-searching
Where do males search for females? Do they aggregate in groups?
Is there some factor that might attract females, such as flowers or egg-laying sites? Do females move to sites where males are present?
How do males search? Do they change strategy? When do they search?
Are females chased/courted at all times? Do females ever reject male advances? When?
Do males make mistakes in chasing objects? What percentage of chases are after females?
Try marking individual males: do they remain in one area over several days?
Is there evidence for territoriality? Do males fight? If so, does the resident always win? Can you distinguish between male-male and male-female interactions?

Courtship
Is there a definite courtship? What is it like? How long does it last?
What percentage of courtships end in actual mating?
Do other males interrupt courtship?
Do females appear to respond to male advances? How?

Mating
How long does mating last? How does it take place? What are the positions of male and female when mating? Can a male be displaced by another?
Does egg-laying occur straight away?

Post-mating behaviour
Do males guard females until they lay their eggs? How long does the male guard? Can he be displaced by another male?

PLATE 1

1
Syrphus vitripennis
(a) male, (b) female

2
Syrphus ribesii

3
Dasysyrphus albostriatus
(a) male, (b) female

4
Dasysyrphus tricinctus

5
Dasysyrphus venustus

6
Sphaerophoria scripta
(a) male, (b) female

7
Epistrophe eligans
(a) male, (b) female

8
Scaeva pyrastri

9
Metasyrphus luniger
(a) male, (b) female

10
Metasyrphus corollae
(a) male, (b) female

(×2.25 natural size)

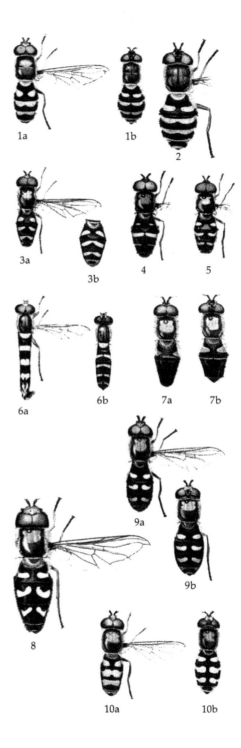

PLATE 2

1
Episyrphus balteatus

2
Leucozona lucorum
(a) typical form, (b) dark

3
Meliscaeva cinctella

4
Meliscaeva auricollis
(a) male, (b) female

5
Baccha obscuripennis

6
Melanostoma scalare
(a) male, (b) female

7
Melanostoma mellinum
(a) male, (b) female

8
Platycheirus albimanus
(a) male, (b) female

9
Platycheirus scutatus
(a) male, (b) female

10
Platycheirus manicatus

11
Platycheirus peltatus
male

12
Platycheirus clypeatus
(a) male, (b) female

13
Chrysotoxum cautum

(×2.25 natural size)

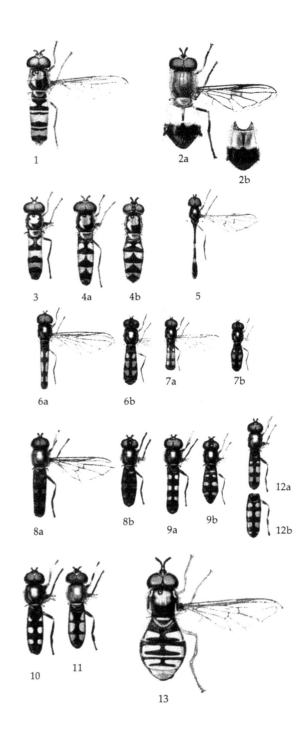

PLATE 3

1
Helophilus pendulus

2
Ferdinandea cuprea

3
Rhingia campestris

4
Xylota segnis

5
Syritta pipiens
(a) male, (b) female

6
Xylota sylvarum

7
Merodon equestris
(a)–(c) colour varieties, all
males

8
Neoascia tenur
(a) male, (b) female

9
Neoascia podagrica
(a) male, (b) female

10
Chrysogaster hirtella
Many species of the genera
Chrysogaster and *Cheilosia*
look very similar; *Cheilosia
paganus* is a small black
hoverfly like the picture, but
with partly yellow legs and
orange antennae. Check
carefully with the descriptions

(1–7 are ×2.25 natural size,
8–10 are ×2.5 natural size)

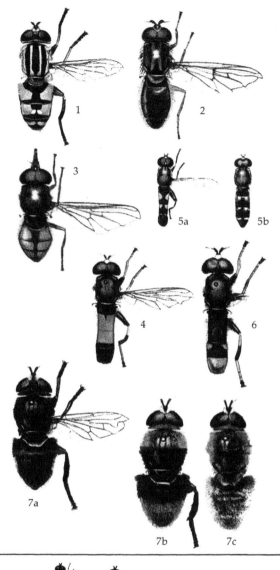

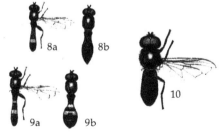

1
Eristalis tenax
(a) dark male, (b) pale male,
(c) pale female

2
Eristalis pertinax

3
Eristalis arbustorum
(a) dark male, (b) pale male

4
Eristalis intricarius
(a) and (b) male, (c) female

5
Myiatropa florea

6
Sericomyia silentis

7
Volucella bombylans
(a) var. *plumata*, (b) typical
form

8
Volucella pellucens

(×1.75 natural size)

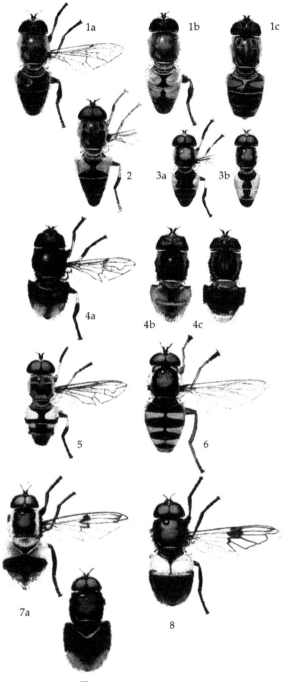

3.6.3 Laying eggs

Having fed on pollen, and having mated, females complete egg development and then seek out places to lay. Most species lay in or at the edge of potential larval food, often being guided by odour. Thus female *Eristalis tenax* are attracted to the smell of dung and stagnant water, and female *Metasyrphus corollae* by odours from aphid honeydew. As a result, egg-batches are judiciously placed in the habitat. Some species are rather more cunning: the larvae of *Rhingia campestris* live in cowpats, and females lay on leaves overhanging the dung. On hatching, the larvae drop straight into their food.

Egg-laying behaviour has been studied most extensively in species with aphid-feeding larvae. Table 6 gives an idea of the sequence of influences that affect egg–laying in these species. These influences are more or less important according to the species involved.
In *Metasyrphus corollae*, for example, the plant-related factors are relatively unimportant, the aphids providing the dominant stimulus to lay. Such eggs as are laid on uninfested plants normally fail to hatch, but egg-laying usually stops in the absence of aphids. Egg development continues in the absence of aphids, though, and mature eggs accumulate in the ovaries. Prolonged retention of eggs leads to their reabsorption, which can be recognised on dissection from the empty egg 'shells' in the ovaries. Nutrients

Table 6. *Influences on egg-laying behaviour (modified from Chandler, 1966)*

Sense involved	Influences
Sight	1. Size of plant patch
	2. Density of plant patch
	3. Colour of plants
	4. Form of plants
Smell	5. Smell of plants
	6. Smell of aphids
Taste	7. Honeydew
Sight	8. Size and position of colony
	9. Shape of aphid
	10. Movement of aphid
Touch	11. Actual site of egg

Response by female	Influences involved
Habitat selection	1,2,3
Plant selection	2,3,4,5
Aphid colony selection	6,7,8,9,10
Egg-site selection	8,9,10,11

reabsorbed from such eggs lengthen the female's life, but overall the total number of eggs that she will lay is reduced. Egg-laying behaviour is similar in *Episyrphus balteatus* and *Scaeva pyrastri*, and probably in other species too.

In other species aphid-related factors are relatively less important, and plants provide an equally important stimulus to lay. These species (*Melanostoma* species and all *Platycheirus* species except *P. scutatus*) lay almost as many eggs on plants without aphids as on those with aphid colonies. In addition, eggs are laid in batches of two to four, instead of singly. The net effect is that these species exploit small aphid colonies that do not attract species such as *Syrphus ribesii* and *Metasyrphus corollae*. The first larva to hatch cannibalises the others, and then searches for aphids. This pattern has obvious relevance for the control of aphids, since *Platycheirus* and *Melanostoma* will lay in advance of aphid attack, and if aphids do not appear, the larvae will not harm the crop.

Female age is a complicating factor in this scheme. Experiments show that a young female *Episyrphus balteatus* lays only on plants with aphids. As she ages, however, discrimination is lost and she will lay without aphids.

There are some organisms, usually mites, that take advantage of the egg-laying habits of hoverflies to move themselves between patches of their own food. For example, one species of mite lives in sap flows of elm trees. These sap flows are also the larval habitat (and adult food) of a genus of hoverfly (*Brachyopa*: not in this book) that lays in the spring. The mite enters a dispersal phase in the spring, and grasps the body hairs of hoverflies visiting the sap flows, thereby being transported to another sap flow as a hitch-hiker. There are several examples of this phenomenon. Perhaps mites are most likely to be encountered on hoverflies that lay their eggs in rotting vegetable matter, where mites are abundant: *Syritta* is commonly found with such hitch-hikers.

3.6.4 *Death*

It is unlikely that many adults die of old age. Unlike larvae, though, they are rarely parasitised by other insects. Nematodes, apparently harmless, are sometimes found in the gut or body cavity, especially in *Xylota* . You may also come across the blackened Malpighian tubules (see fig. 67, p. 58) characteristic of a particular fungal infection. Fatal fungal attacks are very common among grassland hoverflies. *Melanostoma* species and *Platycheirus clypeatus* are peculiarly susceptible, and dead and dying individuals are often found attached to the flowering heads of grasses, their bodies seemingly 'blown apart' by expansion of the fungus. The fungus looks like a whitish 'fur' between the abdominal segments. Most of these infections are caused by fungi of the genera *Entomophthora* or *Empusa*.

The major cause of death among adults of other hoverfly species is probably being eaten by other animals. A wide variety of predators has been recorded, and there are also many instances of hoverflies being killed and digested by plants such as sundews and pitcher plants. One plant, a type of water lily, drowns hoverflies in its flowers and is pollinated by pollen sticking to their bodies.

Other insects and spiders are probably the main predators. Several solitary wasps, such as *Ectemnius cavifrons*, specialise in taking hoverflies (see Pickard, 1975). If you can find the burrows of these wasps (see Yeo & Corbet, 1983) it might be possible to carry out a detailed study of their success rate and prey selectivity. By digging out the burrows and identifying the prey, the frequency of prey species can be compared with their abundance in the surrounding habitat. Some degree of selectivity will probably be found. In many studies of this nature the wasps have been found to catch only male flies. Perhaps they find sites where males search for females: this would constitute an important risk that males run when finding a mate. (In the interests of conservation, burrows of solitary wasps should not be excavated unless they are destined for destruction in any case, for instance if standing dead wood is to be cleared.)

Social wasps also take hoverflies as food. In late summer during years of high wasp abundance, flowers can be littered with the results of their attentions, with the wings, legs, heads and abdomens of hoverflies lying scattered around the blossoms. Not all these corpses are necessarily due to wasps though. Some may be due to attacks by large rapacious flies such as dung flies *Scathophaga stercoraria*.

The remaining group of predators is birds, principally flycatchers and other specialised insect-feeders such as swifts and swallows. It is unusual for birds to confine their attentions to hoverflies, probably because these flies are not as abundant as other prey, or are more difficult to catch, and therefore are not as profitable in energy terms.

Of what use, then, are the bright colours of many hoverflies? Classically they are supposed to mimic bees and wasps, and so protect the flies from being eaten as potential predators try to avoid being stung. Some experiments using toads as predators showed that *Eristalis* were protected by their resemblance to honeybees. However, only certain hoverflies are convincing mimics to the human eye, and most of the black-and-yellow species (Syrphinae) seem rather poor wasp mimics. At least one bird can tell hoverfly from wasp as easily as we can. Davies (1977) fed both real wasps and these poor mimics to a tame Spotted Flycatcher.

The bird vigorously rubbed off the stings of wasps before eating them, but took no such precautions with the hoverflies. Clearly there are problems in regarding many hoverflies as mimics.

However, some recent fascinating experiments have been done to find out just what birds think of hoverfly colour patterns (Dittrich and others, 1993). By using special training techniques developed by psychologists, researchers have been able to see how closely different hoverfly patterns were judged by pigeons to resemble wasps. In general the order of similarity to wasps was similar to what we ourselves think, but there were two interesting exceptions. The patterns considered by the pigeons to be the most wasp-like of all were *Episyrphus balteatus* and *Syrphus ribesii* ! These are not at all what we would think of as very wasp-like. Thus we have to be careful not to assume too much about how birds view their insect prey.

There is a good deal to learn about how hoverfly patterns evolved; the mimetic patterns have clearly evolved several times independently during hoverfly evolution. We still know virtually nothing about predation in the field, especially by birds - yet another gap in knowledge waiting to be filled!

4 Identifying hoverflies

4.1 Introduction

Fig. 15. The eyes touch at the top in most male hoverflies.

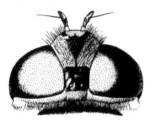

Fig. 16. The eyes are separated in most female hoverflies.

Fig. 17. Male genitalia.

Fig. 18. Female genitalia.

Methods for catching and preserving hoverflies are outlined in Section 5.1. It is better not to put specimens immediately on a pin: dried specimens are more difficult to identify because you cannot move the legs and wings to get the best view of the various characters. Instead, put specimens in the freezer compartment of a refrigerator after capture. They can then be identified, pinned to make a collection, or dissected as part of an ecological study. Specimens for dissection should not be left in the freezer for more than a few days, though, because the internal organs deteriorate unless the insect is frozen at a really low temperature (for example in a deep-freeze). A reference collection should always be made when conducting an ecological study. The collection can be deposited at a local museum, and reference made to it in the publication that results from the work. This ensures that identifications can be checked if necessary.

4.2 How to tell the sexes apart

Male and female hoverflies are easily distinguished. In nearly all species the eyes of the male touch at the top of the head (fig. 15), whereas the eyes of the female are separated by a gap (fig. 16). In some species (for instance, *Helophilus* species), both sexes have the eyes separated. An infallible way of telling the sexes apart is to look at the underside of the tip of the abdomen: males have curved, asymmetrical genitalia (fig. 17) whereas females have pointed abdomens with unobtrusive genitalia (fig. 18).

4.3 How to identify specimens

There are some 250 species of hoverfly in Britain. It will not be long, therefore, before you find species that are not in this book. The object of this chapter is to enable you to name the common species, *and* to say when you have a species not included here.

To identify a specimen, find the species most like it by comparing it with the colour plates. Do not forget to check the pattern of veins in the wings - often a good indicator of the genus. If your specimen is completely different from any of the plates then it is not included in the book, and you will have to try a comprehensive text (Stubbs & Falk, 1983, has pictures and keys to all species) or ask someone familiar with hoverflies.

Fig. 19. Parts of the hoverfly head.

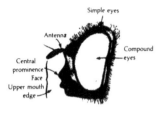

Next, turn to the description of the species you think it may be, and check the specimen against *all* the characters listed under the appropriate species heading. If you cannot get a match with the characters listed, try another species that looks similar. If this still fails to match, then you cannot be certain of the identity of the specimen: try using the pictures and keys in Stubbs & Falk (1983). Nearly all specimens should be identifiable. Be very careful to distinguish between the species you think it may be, and those listed under 'similar species'. The technical terms used in the descriptions are displayed in figs 19-23.

Fig. 20. Thorax from the side.

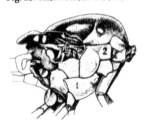

Finally, turn to the description of the genus to which the tentatively identified specimen belongs, and check characters again. **Pay particular attention to the characters listed in italics; these are the crucial ones**. If the picture is very similar to your specimen, and the characters given under the species are correct, it is unlikely that the generic characters will disagree. However, check carefully with those genera listed as 'similar'. If the characters really do disagree, then the specimen remains unidentified. Be careful to distinguish between *Syrphus* and *Epistrophe* (presence or absence of long hairs on the squama: figs 24 and 27), and between *Melanostoma* and *Platycheirus* (the shape of a cuticular plate: figs 32 and 33). Even a moderate familiarity with these genera will render the distinctions very straightforward.

Fig. 21. The hind leg.

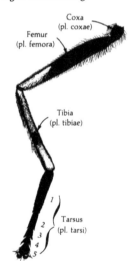

If your specimen closely resembles one of the plates, and the characters all match, you can be reasonably confident of its identity. In practice, nearly all specimens will be identifiable using this book, and nearly all species are distinctive enough for there to be little doubt about the identification. Only certain pairs of species and genera may

Fig. 22. The body of a typical hoverfly.

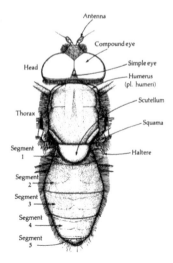

Fig. 23. The antenna.

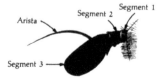

cause problems: in particular, beware of small entirely black specimens. If the generic and tribal characters agree, but the specific ones do not, then the specimen can be labelled to genus. Ecological work is still valid with organisms identified to genus, but will have more value if you use the comprehensive keys in Stubbs & Falk (1983) to find the species.

It would be good practice to check your identifications against the superbly detailed descriptions in Verrall (1901, reprinted 1969), at least until you familiarise yourself with the common species. Verrall's book can be obtained from a library, or from the publishers, E. W. Classey (see Appendix). Do not use Verrall's keys, as these are badly out of date. The names of many of the genera have changed, but nearly all the specific names remain the same (see Stubbs & Falk, 1983).

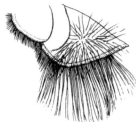

Fig. 24. Long hairs on the squama of *Syrphus* (compare fig. 27).

Once the species is identified, you may like to look for the characters that identify it as belonging to the relevant tribe or subfamily.

In the descriptions, all characters contained in a sentence beginning with '*male*' or '*female*' refer only to that sex. Under each species is also listed the time of year that you can expect to encounter adults. Under the 'similar' species or genera, the numbers or letters refer you to the relevant species or genera that you should check.

4.4 CHARACTERS TO MATCH WITH THE SPECIMEN

SUBFAMILY SYRPHINAE
Humeri bare, without hairs. Back of head hollowed, and closely pressed against the thorax so that the humeri are partly or entirely hidden. Face with at least a weak central prominence (fig. 19), and never flat or concave. Cross-vein 1 before middle of cell B (fig. 3) . Hind femora without strong spines underneath. About 23 genera in Britain.

Fig. 25. Abdomen of *Metasyrphus*.

'Ridge'

Tribe Syrphini
Face and/or scutellum entirely or partly yellow or yellowish-brown. Antennae of normal shape (fig. 23), except in *Chrysotoxum* (J). About 18 genera in Britain.

A. Genus *Syrphus*
Squama with long pale hairs on upper surface, as well as along edge (fig. 24). Abdomen with a ridge at the sides (as in fig. 25). Face yellow. Wing as in fig. 3. Three British species. Similar genera: *Epistrophe* (B); *Parasyrphus* (squama without long hairs on upper surface).

1. *Syrphus ribesii* (L.) (pl. 1.2)
Eyes *bare*, or with *only a few scattered short hairs*. Wing cell E
(fig. 3) *completely covered* with tiny hairs. Thorax dull bronze,
with faint traces of narrow dark lines. *Male*, legs mostly
orange, with black bases; hind femora at least half black.
Female, legs entirely orange except extreme bases and tarsi.
April to November.

Similar species: S. *torvus* (hairy eyes); S. *vitripennis* (2).

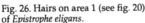
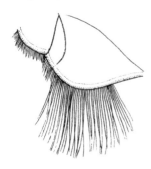
Fig. 26. Hairs on area 1 (see fig. 20) of *Epistrophe eligans*.

2. *Syrphus vitripennis* Meigen (pl. 1.1)
Eyes *bare*, or with only a few scattered very short hairs.
Wing cell E (fig. 3) *with a patch devoid of tiny hairs*. *Male*, hind
femora usually mostly black. *Female*, hind femora nearly all
black.
End of March to November.
Similar species: S. *torvus* (hairy eyes); S. *ribesii* (1).

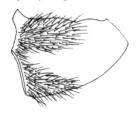
Fig. 27. Bare squama of *Epistrophe* (compare fig. 24).

B. Genus *Epistrophe*
Squama with *no long hairs* on upper surface (fig. 27). Eyes
bare. Area 1 (fig. 20) with two patches of hairs well
separated, but normally *joined by a narrow band of hairs* at the
front edge (fig. 26). Face usually yellow, dusted, sometimes
brownish on the midline. Top of thorax black, sometimes
with stripes, shining, usually dusted yellow at the sides.
Sides of abdomen without the ridge of fig. 25. Five British
species.
Similar genera: *Syrphus* (A); *Parasyrphus* (area 2 with long
hairs).

3. *Epistrophe eligans* (Harris) (pl. 1.7)
Abdominal pattern distinctive. Face completely orange,
obscured by dust and pale hairs. Third antennal segment
dark, more or less orange underneath. Thorax with tawny
hairs.
End of April to June, sometimes into August.

C. Genus *Metasyrphus*
Eyes *bare*. Side of abdomen *with a well-defined ridge* (fig. 25).
Cuticular plate between bases of second and third pair of
legs (compare with fig. 33) normally *with long hairs*. Most or
all of wing covered with tiny hairs. Face usually with a brown
or black stripe down the midline. Six British species.
Similar genera: *Dasysyrphus* (E); *Scaeva* (D).

4. *Metasyrphus corollae* (Fabricius) (pl. 1.10)
Females are difficult to distinguish from several much rarer
species of the genus; it would be wise to check against other
keys. Face yellow, with a blackish line running up from the
mouth edge. Hairs between the antennae and simple eyes

are black. Scutellar hairs yellow. Vein *b* (fig. 3) straight, never dipped. *Male, genital segments remarkably large and conspicuous. Female,* the area between the antennae and simple eyes is *black for about the rear one-quarter to one-third,* the rest being yellow. Abdominal spots reach the side margins.
April to October.
Similar species: *M. luniger* (5); other *Metasyrphus* (male, genital segments normal; female, area between antennae and simple eyes black for one-half or more).

5. *Metasyrphus luniger* (Meigen) (pl. 1.9)
Difficult to distinguish from several much rarer species of the genus; it would be wise to check against other keys. *Male,* alula (fig. 3) *with a bare patch, devoid of tiny hairs;* angle between eyes at top of head *distinctly less than 90°;* abdominal margin more black than yellow, or completely black. *Female,* rear *half or more* of area between simple eyes and antennae is black (ignore the *large gold dust spots* that spread so that only *one-third* of distance between eyes is exposed); abdominal spots do not reach side margins; front femora with *pale* hairs behind.
April to November.
Similar species: *M. corollae* (4); other *Metasyrphus* (alula completely covered with tiny hairs, or (males) angle between eyes more than 90°, or (females) small gold dust spots, or not corresponding to the above description).

D. Genus *Scaeva*
Eyes densely hairy. Most of wing without tiny hairs. Large species. Area between antennae and simple eyes is swollen, especially in males (fig. 28). Face swollen, yellow with a dark stripe along the midline. Top of thorax shining black. Four British species.
Similar genera: *Metasyrphus* (C); *Dasysyrphus* (E).

Fig. 28. Male *Scaeva pyrastri.*

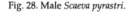

Swollen area

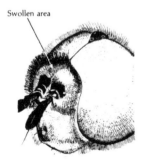

6. *Scaeva pyrastri* (L.) (pl. 1.8)
Inner ends of whitish-yellow abdominal markings on segments 3 and 4 are in a *much more forward position than the outer ends.* Outer ends finish at the middle of the segment, where they occupy little more than one-third of the length of the segment.
May to November.
Similar species: *Scaeva selenitica* (fig. 1) (inner ends of abdominal markings at about the same level as the outer ends).

E. Genus *Dasysyrphus*
Eyes densely hairy. Abdomen *with ridge along sides* (fig. 25).
Cuticular plate between second and third pair of legs
(compare with fig. 33) *not bearing long hairs.* Stigma (fig. 3)
black. Face yellow, normally with a black stripe along the
midline. Five British species.
Similar genera: *Metasyrphus* (C); *Scaeva* (D).

7. Dasysyrphus albostriatus (Fallén) (pl. 1.3)
Two distinct greyish dust stripes at the front half of the thorax.
Stigma (fig. 3) often black. Third and fourth abdominal
segments each with a pair of straight yellow bars that slope
obliquely towards the side margins.
May to October.

8. Dasysyrphus tricinctus (Fallén) (pl. 1.4)
Thorax *unmarked,* shining black. Second abdominal segment
with or without a pair of narrow isolated spots. Third
segment with a broad yellow band, divided or almost
divided into spots. Fourth segment with a yellow basal
band, entire or divided, that is at its broadest *not much more
than one-half as broad* as the band on the third segment.
April to September.

9. Dasysyrphus venustus (Meigen) (pl. 1.5)
Thorax *unmarked,* shining black. Antennae brownish-red to
orange. Scutellum usually pale-haired. Spots on fourth
abdominal segment at broadest *much more than one-half as
broad* as those on the third segment. Spots on the third and
fourth abdominal segments *reach the side margins.*
April to June.
Similar species: Other *Dasysyrphus* (spots do not reach side
margins, or antennae very dark to black).

F. Genus *Leucozona*
Eyes hairy. Face dusted, entirely yellow, or with a dark stripe
along the midline: *L. lucorum is* distinctive. One British
species.
Similar genus: *Eriozona* (some abdominal hairs red; stigma
pale).

10. Leucozona lucorum (L.) (pl. 2.2)
Stigma black. Body rather densely hairy, abdominal hairs
yellow, white, and black, but not red. Face heavily dusted
with yellow, except on the broad shining black stripe along
the midline. Antennae black. Femora black with yellow tips.
May to August.

Fig. 29. Long hairs on area 2 (see fig. 20) of *Meliscaeva*.

G. Genus *Meliscaeva*
Eyes *bare*. Area 2 *with a few long hairs* (fig. 29), sometimes difficult to see. Cuticular plate between second and third pair of legs (compare with fig. 33) *without long hairs*. Extreme rear edge of wing consists of *a series of minute, closely spaced black dots* (use a microscope). Two British species.
Similar genus: *Melangyna* (rear edge of wing without dots).

11. *Meliscaeva auricollis* (Meigen) (pl. 2.4)
Area immediately above antennae *orange*. Third and fourth abdominal segments with yellow (sometimes partly or entirely metallic grey) bands, *deeply cut away* at the middle of the rear edge, *often giving separate spots*. Face and legs variable in colour.
February to November.
Similar species: *M. cinctella* (12).

12. *Meliscaeva cinctella* (Zetterstedt) (pl. 2.3)
Area immediately above antennae *black*. Third and fourth abdominal segments with straight yellow bands *not, or only moderately, cut away* at the middle of the rear edge, and never giving separate spots. Face yellow, sometimes dark on the central prominence. Hind femora black except at the extreme base and tip.
April to October.
Similar species: *M. auricollis* (11).

H. Genus *Episyrphus*
Area 2 with some long hairs (fig. 29), sometimes difficult to see. *Abdominal pattern unique and unmistakable.* Cuticular plate between second and third pair of legs (compare with fig. 33) with long hairs.

Fig. 30. Genitalia of male *Sphaerophoria*.

13. *Episyrphus balteatus* (De Geer) (pl. 2.1)
Only one, unmistakable, species: use generic characters.
February to November.

I. Genus *Sphaerophoria*
Sides of top part of thorax *with sharply defined clear yellow stripes*. Abdomen *without a ridge* at sides, unlike fig. 25. Fringe of hairs underneath edge of scutellum (fig. 43, p. 00) *with at least a large gap in the middle, and often without any hairs at all*. Cuticular plate between second and third pair of legs (compare with fig. 33) *with long hairs*. Male genital structures *very large*, forming a globular tip to the abdomen (fig. 30). About nine British species.

14. *Sphaerophoria scripta* (L.) (pl. 1.6)
Males identifiable with certainty, but *females are not*. Thorax
with a *continuous yellow stripe* at sides, extending from
humerus to scutellum. Hind femora underneath on the top
half *with an area of black bristles* larger than those covering the
rest of the femora. *Male*, abdomen *strikingly long*, obviously
longer than wings. *Female, not* separable from other species
in the genus: call them *Sphaerophoria* sp. unless a mating pair
is caught.
May to October.
Similar species: Other *Sphaerophoria* (males, yellow thoracic
stripe extends only from humerus to furrow at middle of
thorax, or abdomen not longer than wings).

Fig. 31 Antenna of *Chrysotoxum*.

J. Genus *Chrysotoxum*
Antennae *very long*, held pointed forwards (fig. 31).
Abdomen *strongly convex*, with a strong ridge at sides.
Distinctive abdominal coloration. Eight British species.

15. *Chrysotoxum cautum* (Harris) (pl. 2.13)
Abdominal edges with the ridge *partly or entirely yellow*.
Stigma (fig. 3) *orange*. Third antennal segment *only two-
thirds as long* as the second; the second antennal segment
slightly shorter than the first. *Male*, genitalia *very large*.
May to August.
Similar species: Other *Chrysotoxum* (abdominal ridge all
black, or antennal ratios different).

Tribe Bacchini
Face and scutellum *entirely black* in ground colour. Abdomen
without a ridge at sides, unlike fig. 25. Cuticular area
between second and third pair of legs (compare with fig. 33)
without long hairs. Four British genera.

Fig. 32. Underneath the thorax-
abdomen junction of *Melanostoma*.

Note shape of this plate

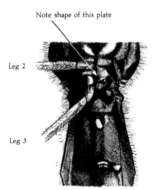

Leg 2

Leg 3

K. Genus *Baccha*
Abdomen stalked (pl. 2.5). Alula of wing (fig. 3) *almost
absent*. Eyes bare. Unmistakably long and thin species. Some
authors consider that there are two species; others think
there is only one, and call it *B. elongata*.

16. *Baccha obscuripennis* Meigen (pl. 2.5)
Unmistakable species: use generic characters.
May to October.

L. Genus *Melanostoma*
Cuticular plate between second and third pair of legs
reduced to a spear-shaped band (fig. 32). Front tibiae and tarsi
of males not noticeably flattened. It is common to find

Fig. 33. Underneath the thorax-abdomen junction of *Platycheirus*.

Note shape of this plate

Leg 2

Leg 3

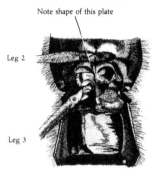

Fig. 34. Front leg of male
Platycheirus albimanus.

Fig. 35. Profile of female
Platycheirus albimanus.

entirely black variants of these species. Three British species. Similar genus: *Platycheirus* (M).

17. *Melanostoma mellinum* (L.) (pl. 2.7)
Male, second abdominal plate on underside of abdomen *is at most one-and-a-half times* as long as wide. **Female**, dust spots between antennae and simple eyes *restricted to a thin area at sides*, occupying at most *no more than one-quarter* of the distance between the eyes.
April to November.
Similar species: *M. scalare* (18).

18. *Melanostoma scalare* (Fabricius) (pl. 2.6)
Male, second abdominal plate on underside of abdomen *is twice as long as wide*, and is *extensively pale*. **Female**, *sharply defined large dust spots* between the antennae and simple eyes, occupying *one-half or more* of the distance between the eyes.
April to November.
Similar species: *M. mellinum* (17)

M. Genus *Platycheirus*
Cuticular plate between second and third pair of legs is of *normal shape* (fig. 33). Front tibiae and/or tarsi of males *often conspicuously flattened*. This can be a difficult genus to identify, particularly the females. It would be wise to take specimens through the more comprehensive keys to get used to the characters, before relying entirely on those given here. Eighteen British species.
Similar genera: *Melanostoma* (L); some Syrphini (face and/or scutellum not entirely black).

19. *Platycheirus albimanus* (Fabricius) (pl. 2.8)
Male, front tibiae *abruptly enlarged just before the tip* after some long black hairs (fig. 34); front femora *with clumps of tangled hairs* behind near the base (fig. 34); front tarsal segment 1 *up to about three times as long* as the narrower segment 2 (fig. 34); third antennal segment more or less obviously pale underneath. **Female**, thorax *shining;* antennae *partly yellow or completely black;* area between antennae and simple eyes with *inconspicuous* grey side spots *not nearly* occupying two-thirds of the distance between the eyes; face evenly dusted all over, except sometimes on the central prominence which projects further than the upper mouth edge (fig. 35); fore- and mid-legs normally *extensively darkened,* with at least the tibiae darkened after the middle.
April to November.
Similar species: Other *Platycheirus* (including 20, 22, 23).

Fig. 36. Front leg of male
Platycheirus clypeatus.

Fig. 37. Profile of female
Platycheirus clypeatus.

Fig. 38. Profile of *Platycheirus manicatus.*

20. *Platycheirus clypeatus* (Meigen) (pl. 2.12)
Antennae *absolutely black.* Second and third abdominal
segments *at least as wide as long. Male,* front tibiae and tarsi
obviously enlarged, as in fig. 36; front femora *with a tuft of two
or three long white hairs* near the base (they intermingle and
seem like a single hair), then a few long weak *black* hairs
bent forwards at their tips, and then near the tips of the
femora a few shorter weak black hairs; hind femora *mainly
blackish;* fifth abdominal segment largely black; halteres
yellow. *Female,* thorax shining: central prominence projects
further than the upper mouth edge (fig. 37); area between
antennae and simple eyes with conspicuous grey side spots
occupying about two-thirds of the distance between the
eyes; front femora with hairs becoming *increasingly longer*
towards the base; hind femora and tibiae with dark
markings; sixth abdominal segment black. April to October.
Similar species: Other *Platycheirus* (including 19, 22, 23).

Fig. 39. Front leg of male
Platycheirus peltatus.

21. *Platycheirus manicatus* (Meigen) (pl. 2.10)
Thorax *dull greenish-black* with obscure dark markings
running from front to back. Upper mouth edge *greatly
extended forwards* (fig. 38). *Male,* front tibiae only *slightly
enlarged* even at the tip, but front tarsal segment 1 is *large,
white and round.*
May to July, August to October.

22. *Platycheirus peltatus* (Meigen) (pl. 2.11)
Male, front tibiae *abruptly enlarged just before the tip* after
some long black hairs (fig. 39); front femora with *strong,
coarse, bristly hairs behind for entire length* (fig. 39); tarsal
segment 1 of front leg obviously white even without a
microscope, and has a *characteristic shape* (fig. 39); thorax
shining black; third antennal segment more or less
obviously pale underneath. *Female,* thorax *shining* black;
third antennal segment *at least partly pale* underneath; area
between antennae and simple eyes *with conspicuous grey* side
spots occupying *two-thirds or more* of the distance between
the eyes at their widest point; grey side spots *widen
gradually* not abruptly, to their maximum width; sides of

Fig. 40. Profile of female
Platycheirus peltatus.

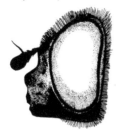

Fig. 41. Front leg of male
Platycheirus scutatus.

thorax heavily dusted grey; lower face and central prominence quite strongly extended forwards (fig. 40). May to October.
Similar species: Other *Platycheirus* (including 19, 20, 23).
23. **Platycheirus scutatus** (Meigen) (pl. 2.9)
Male, front tibiae *abruptly enlarged just before the tip* after some long black hairs (fig. 41); front tarsal segment 1 *about six times as long as the very short but equally wide* segment 2 (fig. 41); coxae of second pair of legs with a long finger-like projection; front femora *with clumps of tangled hairs* behind near the base (fig. 41); third antennal segment more or less obviously pale underneath. *Female,* thorax *shining;* third antennal segment *pale underneath;* area between antennae and simple eyes *with conspicuous grey side spots* occupying *two-thirds or more* of the distance between the eyes at their widest point; grey side spots confined to eye margins, then *abruptly and narrowly* extending across width; sides of thorax behind humeri undusted, or only lightly dusted, and *brightly shining.*
April to October.
Similar species: Other *Platycheirus* (including 19, 20, 22).

SUBFAMILY ERISTALINAE
Humeri with *at least a few, and usually many* hairs. Back of head not strongly concave, and not strongly pressed to thorax; humeri are therefore usually clearly exposed. Forty British genera.

Tribe Cheilosiini
Cross-vein 1 *before the middle of* cell B (as in fig. 3), except in *Ferdinandea,* where cross-vein 1 is *at the middle* (fig. 42).
Arista bare, or with *very* short dense hairs (pubescence). In profile, face always extended forwards. Face always with a central prominence and/or an extended upper mouth edge. Scutellum has *a fringe of hairs underneath the edge* (fig. 43). Four British genera.

Fig. 42. Wing of *Ferdinandea.*

Fig. 43. *The hair fringe under the scutellum.*

Bristles

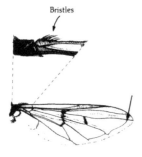

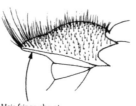

Hair fringe absent
(*Chrysogaster*)

Hair fringe present
(*Cheilosia*)

N. Genus *Cheilosia*
Body normally *black*. Top surface of abdominal segments *shining or partly shining*. Face with a central prominence (see fig. 19). Leg colour variable. This is a very difficult genus, and it would be helpful to consult Stubbs & Falk (1983) to see what to look for. About 33 British species.
Similar genus: *Chrysogaster* (Q).

24. *Cheilosia paganus* (Meigen)
A small black species similar to *Chrysogaster hirtella* (pl. 3.10). Eyes *bare*. Legs *partly yellow*, including the basal part of the tibiae. Face *without* any long erect hairs (as opposed to short, dense hairs (=pubescence)). Third antennal segment *orange*, often with a black tip. Arista *without obvious* hairs or pubescence at the base. Underneath of hind femora *without* long whitish hairs that are longer than the width of the scutellum. *Female*, tip of scutellum *entirely dark*.
Similar species: Other *Cheilosia* (check very carefully the above characters).

Fig. 44. Profile of *Rhingia campestris*.

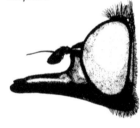

O. Genus *Rhingia*
Unmistakable: face extended forwards into a long snout (fig. 44). Two British species.

25. *Rhingia campestris* Meigen (pl. 3.3)
Sides of abdomen *continuously and broadly* black, with pale hairs. At least the hind tibiae with a blackish ring. Thorax black, lightly dusted.
April to November.
Similar species: *R. rostrata* (sides of abdomen continuously orange).

P. Genus *Ferdinandea*
Face *without* a central prominence. Wing *with conspicuous bristles* along one of the basal veins (fig. 42). Thorax with conspicuous bristles, particularly along sides and on margin of scutellum. Two British species.

26. *Ferdinandea cuprea* (Scopoli) (pl. 3.2)
Abdominal segments *gleaming brassy*. Front and middle femora, and all tibiae, with some strong black bristles.
April to October.
Similar species: *F. ruficornis* (abdomen bluish-black; front and middle legs without or with very few bristles).

Tribe Chrysogasterini
Cross-vein 1 *before middle* of cell B (see fig. 3). Arista bare, or with tiny hairs. In profile, face always extended forwards, and with a central prominence and/or an extended upper

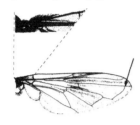

Fig. 45. Wing of *Chrysogaster*.

mouth edge. Hind femora *with spines,* sometimes small. Normally *no fringe of hairs* on the lower edge of the scutellum (fig. 43). Seven British genera.

Q. Genus *Chrysogaster*
Top of abdomen mostly dull or semi-shining black, but edges are shining in all segments *except* the first (whose margins are dull). Upper part of vein *a* joins vein *b* some distance from the tip of the wing (fig. 45). *Male, swollen* area between antennae and simple eyes (fig. 46), eyes touch at top. *Female,* face *without* a central prominence; area between antennae and simple eyes bears transverse furrows, sometimes hard to see; *no stripes on thorax.* Five British species.

Fig. 46. Head of male *Chrysogaster*.   Similar genus: *Cheilosia* (N).

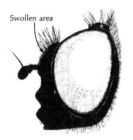

Swollen area

27. *Chrysogaster hirtella* Loew (pl. 3.10)
This is a very common species, but is moderately difficult to separate from other *Chrysogaster* species. Check characters carefully. First abdominal segment (fig. 22) dull on top except sometimes the extreme hind margin towards the sides in females. Area above and behind front leg bases *brightly shining,* undusted. *Male,* top of thorax mostly shining; antennae *dark brown or black;* body colour *normally black,* not bronze; top of thorax with many yellowish hairs mixed with the blackish hairs, especially near the wing bases. *Female,* top of thorax covered with short hairs; wing membrane *without* a brownish tinge or cloud; outward-facing top half of middle femora clothed in pale yellowish hairs.
May to August.
Similar species: *C. maquarti* (male, top of thorax mostly, or usually entirely, black-haired; female, middle femora with black hairs. See Speight, 1980).

R. Genus *Neoascia*
Very small species. Face *without* a central prominence, almost straight in profile. Mouth edge extended forwards. Abdomen with first and second segments *much narrower* than the third segment. Arista bare. Outer backward angle of cell A *almost rectangular,* not rounded (fig. 47). Five British species.
Similar genus: *Sphegina* (outer backward angle of cell A rounded; face strongly concave in profile).

Fig. 47. Wing of *Neoascia tenur*.

28. *Neoascia tenur* (Harris) (pl. 3.8)
Upper and lower parts of vein *a* without dark markings (fig. 47). Third antennal segment decidedly *longer than broad.* Hind femora yellow at the base, then black, *including the tip.* Behind the base of the third pair of legs is a 'bridge' *that is complete* (fig. 49), not incomplete. *Male,* third abdominal

Fig. 48. Wing of *Neoascia podagrica*.

segment with a yellow band or pair of spots, of which at least the front margins *never extend to the side margins*. April to October.
Similar species: *N. podagrica* (29); other *Neoascia* (incomplete 'bridge', although gap is often narrow).

29. Neoascia podagrica (Fabricius) (pl. 3.9)
Upper and lower parts of vein *a* with dark markings (fig. 48). Behind the base of the third pair of legs is a 'bridge' *that is complete* (fig. 49), not incomplete.
April to October.
Similar species: *N. tenur* (28); other *Neoascia* (incomplete 'bridge', although gap is often narrow).

Fig. 49. 'Bridge' behind the hind legs of *Neoascia*.

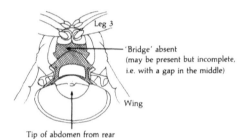

Leg 3

'Bridge' absent
(may be present but incomplete, i.e. with a gap in the middle)

Wing

Tip of abdomen from rear

Fig. 50. Antenna of *Volucella*.

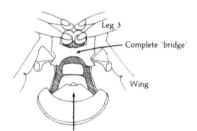

Leg 3

Complete 'bridge'

Wing

Tip of abdomen from rear

Fig. 51. Wing of *Volucella*.

Tribe Volucellini
Arista *feather-like* (fig. 50). Cross-vein 1 before middle of cell B (fig. 51). Upper part of vein *a* bends back strongly towards front of wing (fig. 51). One British genus.

30. Volucella bombylans (L.) (pl. 4.7)
Body densely hairy, *with colour patterns that mimic bumblebees*. No other bumblebee mimic has a feather-like arista (fig. 50).

Fig. 52. Wing of *Sericomyia*.

May to August.
31. ***Volucella pellucens*** (L.) (pl. 4.8)
Body hair inconspicuous. Thorax black, reddish-brown towards sides.
May to September.
Similar species: *V. inflata* (yellow, not white, abdominal markings).

Tribe Sericomyiini

Fig. 53. Wing of *Syritta*.

Arista *feather-like* (fig. 50). Cross-vein 1 *at the middle* of cell B (fig. 52). Upper part of vein *a* bends back moderately towards the front of the wing (fig. 52). One British genus.

T. Genus *Sericomyia*
Use tribal characters. Four British species.

32. ***Sericomyia silentis*** (Harris) (pl. 4.6)

Fig. 54. Hind leg of *Syritta*.

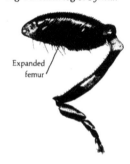

Expanded femur

Body *thinly hairy*. Scutellum usually black. Male genitalia *yellow*. Abdominal stripes *dark* yellow, widening considerably towards sides.
April to September.
Similar species: *S. lappona* (scutellum reddish-brown, abdominal stripes pale yellow or whitish).

Tribe Xylotini
Cross-vein 1 placed *at or beyond* middle of cell B (see fig. 55) (slightly before middle in *Syritta*: fig. 53). Vein *b* never looped as in fig. 58, but *wavy* as in fig. 55. Arista bare. Cross-vein 1 *slanted*, never at right angles between the veins. Thorax without strong bristles. Nine British genera.

Fig. 55. Wing of *Xylota*.

U. Genus *Syritta*
Hind femora *very swollen* (fig. 54). Thorax rather dull black with greyish markings at the front and sides. Scutellum black. Front and mid-legs mainly yellow, tibiae darkened near the tip. Hind femora mainly or partly black, yellow about the base and middle. Tarsi black. One British species.

33. ***Syritta pipiens*** (L.) (pl. 3.5)
Use generic characters.
March to October.

Fig. 56. Head of *Xylota*.

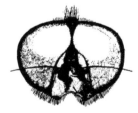

V. Genus *Xylota*
Upper part of vein *a* meets vein *b* *well before the tip* of vein *b*, and is directed upwards at the point of contact (fig. 55). Body hairs short and sparse. Face *not, or scarcely, descending below level of eyes* (fig. 56). Cuticular plate between second and third pair of legs *well developed, and without long hairs* (compare with fig. 33). Viewed from the front, head as in fig.

56, not strongly triangular. Seven British species.
Similar genus: *Chalcosyrphus* (cuticular plate between second
and third pair of legs with long hairs).

34. Xylota segnis (L.) (pl. 3.4)
Hind femora underneath with *two rows* of well-separated,
long, stout spines. Front and mid-legs *partly yellow.* Fourth
abdominal segment with only short hairs.
May to September.
Similar species: *X. tarda* (spines underneath hind femora not
arranged in rows).

Fig. 57. Hind leg of *Merodon equestris.*

Triangular lobe

35. Xylota sylvarum (L.) (pl. 3.6)
Front and mid-legs *partly yellow.* Fourth abdominal segment
clothed in conspicuous, close-set, golden hairs folded down
against the surface of the cuticle. *All tibiae are darkened on the
apical half.*
May to September.
Similar species: *X. xanthocnema* (all tibiae almost entirely
yellow.)

Tribe Eumerini
Area 2 *with long hairs* (as in fig. 29). Eyes and face *hairy.*
Arista bare. Scutellum with a flattened margin, difficult to
see in the very hairy *Merodon.* Characteristically 'hunched'
appearance. The tribe consists of a large bumblebee mimic
(*Merodon*), and a genus of small usually black species with
slanting whitish dust-bars on the abdominal segments
(*Eumerus:* not in this book).

W. Genus *Merodon*

Fig. 58. Wing of *Eristalis.*

Cross-vein 1 placed *at or beyond* middle of cell B. Vein *b
looped* (fig. 58). Upper part of vein *a* bent back towards the
front edge of the wing (as in fig. 51). Hind femora with a
flat, toothed, triangular lobe underneath near the tip (fig. 57).
One British species.

36. Merodon equestris (Fabricius) (pl. 3.7)
Use generic characters. No other bumblebee mimic has a
triangular lobe under the hind femora.
April to August.

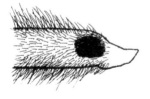

Fig. 59. Base of femur of *Eristalis.*

Tribe Eristalini
Vein *b looped* (fig. 58). Cross-vein 1 placed near to or beyond
middle of cell B (fig. 58). Upper part of vein *a not* bent back
towards front of wing (fig. 58). *All femora with a patch of
small stubby black spines* at the base (fig. 59). Hind femora
without a triangular lobe underneath near the tip, unlike fig.
57. Face with a central prominence. Seven British genera.

Fig. 60. Wing tip of *Helophilus*.

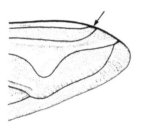

X. Genus *Helophilus*

Cell D *open to the wing margin* (fig. 60). Thorax with grey or yellow stripes. Abdominal segments *without grey stripes running from front to back,* but with some other pattern of grey, yellow or orange markings. Eyes *bare.* Antennae entirely black or dark brown. Eyes in male do not touch at the top of the head. Four British species.

37. *Helophilus pendulus* (L.) (pl. 3.1)

Face *with a black stripe* down the midline. Hind tibiae *yellow on basal two-thirds.*
April to October.
Similar species: Other *Helophilus* (no black facial stripe, or hind tibiae yellow only on basal third).

Fig. 61. Arista of *Eristalis pertinax.*

Y. Genus *Eristalis*

Cell D *closed* near the edge of the wing tip (fig. 58). *Wing as in fig. 58.* Eyes *without* a conspicuous pattern of dark spots. Scutellum yellow, orange or reddish-orange. Nine British species.

38. *Eristalis arbustorum* (L.) (pl. 4.3)

Arista *with long hairs* on basal half *only.* All tarsi more or less darkened. Third antennal segment black or brown. Face *entirely covered with pale dust:* rubbed or old specimens often have a *very narrow* shining stripe along the midline.

Fig. 62. Arista of *Eristalis tenax.*

April to October.
Similar species: Other small *Eristalis* (antennae orange, or arista with only short hairs, or face with a broad black stripe).

39. *Eristalis intricarius* (L.) (pl. 4.4)

An obvious bumblebee mimic, with *long, more or less dense body hair* that may or may not hide the underlying abdominal pattern. Colour of hairs variable.
March to September.

Fig. 63. Head of *Eristalis tenax.*

Similar species: Other bumblebee mimics (wing different, or like *Merodon* (36)).

40. *Eristalis pertinax* (Scopoli) (pl. 4.2)

Arista *as in fig. 61.* Front and mid-tarsi *entirely orange. No stripes on eyes formed by denser rows of hairs, unlike E. tenax* (41). Antennae dark.
March to November.
Similar species: *E. tenax* (41).

41. *Eristalis tenax* (L.) (pl. 4.1)

Arista *with only very short hairs on* basal half (fig. 62). Eyes appear to the naked eye to have *two dark stripes* formed from

rows of denser hairs (fig. 63). All tarsi more or less
darkened. Face with a broad black stripe along the midline.
Throughout the year.
Similar species: *E. pertinax* (40).

Fig. 64. Wing of *Myiatropa*.

Z. Genus *Myiatropa*

Cell D *open to the wing edge* near wing tip (fig. 64). Eyes
hairy. One British species.

42. *Myiatropa florea* (L.) (pl. 4.5)

Use generic characters.
May to October.

5 Techniques

5.1 Collecting

5.1.1 *Catching hoverflies*

Catching can be done with a net, trap, tubes or a pooter. Nets can be simply made, and the best ones can be fitted with a long handle to catch high-flying species: they can also be obtained from entomological suppliers (see Appendix). The best traps for hoverflies are water and Malaise traps. A water trap is simply a coloured dish containing water and a little detergent, placed on the ground or raised up on a stone or stick. White is apparently the most effective colour, with yellow also being good (Disney and others 1982). Unless you intend to check the traps every day, it will be necessary to add to the water 1 cm^3 of a preservative such as formalin (beware, toxic), otherwise the flies will decompose in the trap. Don't add too much or you may repel the flies. Malaise traps (fig. 65) are very effective if positioned correctly, catching enormous numbers. They can be made or bought (see Appendix).

A very instructive and useful project would be to compare the effectiveness of various trapping methods (see Disney and others, 1982). Such a project may shed some light on the position of some species in the community; for example, *Metasyrphus luniger* is very commonly found as a larva in aphid colonies, but is not often caught as an adult.

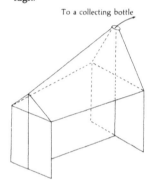

Fig. 65. A Malaise trap, made from gauze, supported with poles and guy ropes like a tent. About 1.5 m high.

To a collecting bottle

5.1.2 *Making a collection*

Techniques for preserving insects, and for making a collection, can be found in Oldroyd's (1958) book. Very little is necessary for collecting hoverflies: a net, small tubes, pins and a suitable box are the only requisites. The box is perhaps the most important item: it needs a cork or expanded polyethylene foam floor to hold the pins, and an airtight lid (see Oldroyd, 1958, for details). A clear plastic sandwich box is suitable.

The best way of killing hoverflies is to put them in tubes in the freezer compartment of a refrigerator. This leaves them relaxed and hence easier to identify, and also preserves them fresh, so that they can be dissected as required. To make a collection, you only need to pin the insects and allow them to dry. Nearly all hoverflies are best pinned by inserting the pin straight down through the top of the thorax, leaving the insect about one-third of the way down from the pin-head. While the fly dries on the pin, the wings and abdomen will need some sort of support,

Polyethylene foam is obtainable as sheets 1000x750x12 mm from Wilford Polyformes, Greaves Way, Stanbridge Road, Leighton Buzzard, Bedfordshire otherwise they will droop. Crossed pins will support the abdomen, and other well-placed pins will keep the wings in the correct position. Each pinned specimen should bear a data label, with at least the date and place of capture.

Using carbon dioxide to anaesthetise field-caught specimens is useful because it preserves, for example, the positions of pollen grains on the body hairs. Carbon dioxide is available in small canisters that fit into a handheld dispenser, normally used for opening wine bottles (address, p. 62).

Larvae are best preserved in 70% alcohol (see Rotheray, 1989a).

5.2 Recording

The importance of recording data cannot be overemphasised. Even if it is only made for the pleasure of looking at the colour patterns of hoverflies, a collection can be useful for ecologists if adequate records are kept of at least the date and place of capture, and preferably the habitat, what flower the fly was on, what it was doing, the weather, and other details.

For an ecological study one must keep records not only of when and where species are seen, but of how many individuals there were, how long they were engaged in various activities, what the temperature was, or other types of *quantitative* data. The decision of what to measure is probably the most crucial one of the entire exercise, and is perhaps one of the areas where advice will be particularly valuable. The decision will be wiser if taken after an adequate period of getting to know the patterns of behaviour of the flies, so that you have an idea of what factors are important to them. For example, in a study of larval behaviour, if the effect of temperature is not of primary interest, the experiments should be conducted at the same temperature to prevent the effects of this variable from interfering with the results of the study.

Many studies can be carried out with a minimum of equipment: a notebook, a stopwatch and a thermometer, perhaps. For electronically minded ecologists, useful advice on building simple equipment is to be found in Unwin's (1980) book on microclimate and its measurement or in Unwin & Corbet (1991).

Often the data that reveal most about animals are those that you can compare. To be comparable, data must be gathered systematically. For instance, regular censusing of feeding hoverflies, each census beginning at the same time of day, covering the same path, and lasting for about the same length of time, can allow a detailed comparison of the diets of different species. Two censuses cannot be compared in a meaningful way if weather patterns were completely different, hence the importance of recording the weather and time of day.

5.3 Marking

Marking techniques are described by Southwood (1978). The most useful marker for hoverflies, both for adults and larvae, is probably the quick-drying enamel paint sold in modelling shops. A spot is applied to the back of the larva, or the top of the thorax of the adult (keeping well clear of the wing bases and the scutellum). Spots of different colours or in different patterns allow individuals to be recognised. Adults are easily marked when anaesthetised, or when immobilised in the net, but they may fly away from the area when released. Males very often return after a few hours, but experience with different releasing techniques may improve the resighting frequency.

5.4 Dissecting hoverflies

It may look difficult to dissect a small hoverfly such as *Melanostoma scalare*, but in fact dissection is very straightforward, especially for smaller species, since the membranes are thin and easily ruptured. The only essential pieces of equipment are a binocular microscope, a pair of fine forceps, and some cutting tools. Some of the smallest entomological pins attached to a holder of some sort will act as suitable scalpels. A pin inserted into the end of a hypodermic needle and fixed with a little wax is entirely adequate. Fine forceps and pins can be obtained from entomological suppliers (see Appendix).

Little extra can be learnt from dissecting larvae, since ecologically relevant features such as whether the larva is parasitised and the passage of food can be seen through the cuticle (see Rotheray, 1989a). Readers interested in the fine details of larval anatomy should refer to the paper by Bhatia (1939).

Anatomical details of the head of adults are complex, and are best left to specialists; little of ecological relevance can be learnt from simple dissection. The thorax contains the huge flight muscles, but very little else: it is worth dissecting one thorax to appreciate the relative size and power of these muscles. Use a razor blade to cut a thorax in half along the midline from back to front. Thoracic muscles are repeated on each side, and lie essentially in two layers, the first of which is exposed by the cut. Removing this layer will expose the underlying set.

Features of ecological interest lie in the abdomen. Internally, most species are similar, the main differences being between the sexes. Individual anatomies may look different because some specimens have full and others empty crops, or some females may have fully developed eggs and others not. Basic anatomical features are shown in figs. 66 and 67.

Fig. 66. Male anatomy.

Fig. 67. Female anatomy.

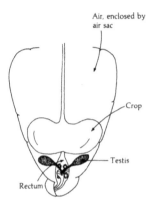

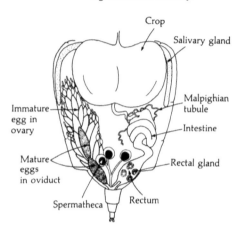

Secure the insect with a pin through the thorax so that the underside is facing up. Prick a hole in the soft cuticle of the abdomen right at the point where thorax and abdomen join. Insert the end of one arm of the forceps into the hole, and grip the tougher cuticle of the first ventral abdominal plate (see fig. 68). Next, using a fine pin slit down both sides of the abdomen right to the tip, following the soft cuticle between the ventral plates and the abdominal margin. Try not to insert the pin far into the abdominal cavity, since you risk puncturing some of the structures. The best instrument for this action is a pin whose end has been curved by tapping it against a hard surface: with this the soft cuticle can be ripped rather easily. All that remains is to peel back the ventral surface of the abdomen,

Fig. 68. How to dissect a hoverfly abdomen.

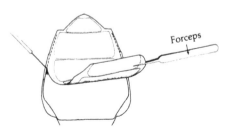

gently freeing it from the underlying membranes (see fig. 68). The crop can be removed in its entirety by grasping the crop duct with forceps, and gently pulling the crop away from the abdominal cavity. Add a drop of water to prevent drying.

It is usually obvious whether the crop or intestine contains any pollen, since most pollen eaten by hoverflies is yellow. If the female has mature eggs, these can in the Syrphini often be seen through the semi-transparent sides of the abdomen before dissection.

Mature eggs are obvious upon dissection because of their large size and the characteristic net-like or ribbed patterning of the surface. Male reproductive organs are less easy to see, and reside in the very tip of the abdomen.

If it is possible to use a good compound microscope, it is a simple process to check whether females have been mated, or whether males are mature (that is, are producing sperm). For females, remove the three dark spermathecae (see fig. 67) and place them in a drop of water on a microscope slide. Place a glass coverslip over the top, and press down to squash the spermathecae. Under high power (about x400) you will be able to see bundles of threadlike sperm if the female has been mated. Note that the absence of sperm may not imply the virginity of the female, unless she has been dissected very fresh. Sperm are apparently highly perishable, and specimens must be dissected as soon as possible after collection to make sure that sperm will be seen if present. A similar squash of male testes will reveal sperm in a mature male. Note that a male may have sperm in the testes and still be incapable of inseminating a female because his accessory glands may as yet be immature.

5.5 Rearing

The easiest way to collect material for rearing is to find females with mature eggs. On capture, if presented with suitable larval food, they will usually lay in small containers. Alternatively, you can search for the larvae themselves. Aphid-feeding species can be found wherever aphids are found. A good method is to bring indoors plants with aphids on them: larvae that hide during the day often emerge at night to feed. Putting collected plants into plastic bags will ensure high humidity, and the larvae will move on to the plastic during the night. Species that do not feed on aphids are generally harder to find, but a search of suitable larval habitats (see table 2) should prove fruitful. All eggs, larvae and pupae must be protected from drying out: placing moist blotting paper in the rearing tubes will overcome this problem. See Rotheray (1989*a*) for further details.

Aphid-feeding species. Record the aphids that the larvae were found on (see Blackman, 1974; Rotheray, 1989*a*), and always try to ensure that there are more aphids provided than the larvae will eat during the night. A large number of aphids will be needed. Provide sand or soil for pupation. Rotheray (1989*a*) gives more details on rearing techniques.

Aquatic species. Keep the larvae in shallow, covered dishes, such as petri dishes, with the liquid in which they were found. Sink the dish in sand or soil, so that the mature larva can climb out and pupate. Keep the water level topped up, and replace if the surface becomes coated with a film of bacteria. *Helophilus* and *Eristalis* have been reared by feeding them on a submerged decaying mouse, or on moist rabbit droppings.

It is a challenging task to try to get adults emerging from your cultures to mate and lay eggs. The conditions have to be varied to find the correct combination. Successful attempts have used high light intensity, a simulated 16-hour day, and, above all, adequate supplies of pollen for food. Many species appear to mate in the air, and need rather large cages (for example 1 m x 1 m x 1 m).

5.6 Identification of rarer species

Identification of species not in this book can be accomplished in three ways. The book by Stubbs & Falk (1983) has fairly simple keys, coupled with illustrations of nearly all species. The older key, by Coe (1953), is not straightforward except for those experienced in hoverfly identification, and it will need supplementing with more modern papers for certain groups. The third alternative is to get someone else to identify specimens. You could join the Diptera Recording Scheme (see Appendix) and go to field meetings, where experts will be able to help you identify unusual species.

5.7 Writing papers

Writing up is an important part of a research project, particularly when the findings are to be communicated to other people. A really thorough, critical investigation that has established new information of general interest may be worth publishing, if the animals upon which it is based can be identified with certainty. In many cases, identification to genus is adequate. Journals that publish short papers on insect biology include the *Entomologist's Monthly Magazine,* the *Entomologist's Record and Journal of Variation, Bulletin of the Amateur Entomologist's Society,* and, for material with an educational slant, the *Journal of Biological Education.* Those unfamiliar with publishing conventions are advised to examine current numbers of these journals to see what sorts of things they publish, and then to write a paper along similar lines, keeping it as short as is consistent with the presentation of enough material to establish the conclusion. It is then time to consult an appropriate expert for advice on whether and in what form the paper might be published. It is an unbreakable convention of scientific publication that results are reported with scrupulous honesty. Hence it is essential to keep detailed and accurate records throughout the investigation, and to distinguish between certainty and probability, and between deduction and speculation. In many cases it will be necessary to apply appropriate statistical techniques to test the significance of the findings. A book such as Chalmers & Parker's (1989) *The OU Project Guide* will help, but this is an area where expert advice can contribute to the planning, as well as the analysis, of the work.

Appendix: some useful addresses

Entomological suppliers

For general entomological equipment:
Watkins & Doncaster, Four Throws, Hawkhurst, Kent.
For CO$_2$ dispensers:
Customer Services Department, Edme Ltd, Mistley,
Manningtree, Essex CO1L 1HG
For Malaise traps:
Marris House Nets Ltd, 54 Richmond Park Avenue,
Queens Park, Bournemouth BH8 9DR.

Societies

Royal Entomological Society, 41 Queen's Gate, London SW7 5HU.
Amateur Entomologists' Society, Registrar, 22 Salisbury Road,
Feltham, Middlesex TW13 5DP.
British Entomological and Natural History Society,
c/o Institute of Biology, 20 Queensberry Place, London SW7 2DZ

Recording hoverfly distributions

Many dipterists take part in recording, so that species distributions
can be mapped. This can be a very worth-while method of getting
acquainted with hoverflies and also with other dipterists. The
address of the coordinator for the Diptera Recording Schemes, and
the coordinator of the Hoverfly Recording Scheme Section, can be
obtained from: Biological Records Centre, Institute of Terrestrial
Ecology, Monks Wood Experimental Station, Abbots Ripton,
Huntingdon PE17 2LS.

Supplier of entomological books, new and second-hand:

E. W. Classey Ltd, PO Box 93, Faringdon, Oxon SN7 7DR.

References and further reading

Finding book and journal articles

Some of the books and journals listed here will be unavailable in local or school libraries. It is possible to make arrangements to see or borrow such works by seeking permission to visit the library of a local university, or by asking your local library to borrow the work (or a photocopy of it) for you via the British Library, Document Supply Centre. Borrowing via the British Library may take several weeks, and it is important to present your librarian with a reference that is correct in every detail. References are acceptable in the form given in the list below, namely the author's name and the date of publication, followed by (for a book) the title and publisher or (for a journal) the title of the article, the journal title, the volume number, and the first and last pages of the article .

General reading

General ecology: Cloudsley-Thompson (1967), Elton (1966), Phillipson (1966), Ricklefs (1991)
Physiology: Heinrich (1979), Schmidt-Nielsen (1975)
Behaviour: Carthy (1979), Heinrich (1979), Krebs & Davies (1987), Thornhill & Alcock (1983)
Insects and plants: Davis (1991), Edwards & Wratten (1980), Redfern (1983), Redfern & Askew (1992), Unwin & Corbet (1991)
Pollination: Proctor & Yeo (1973)
Insect biology: Imms (1971)
Aphids: Blackman (1974), Dixon (1973), Rotheray (1989a)
Techniques: Begon (1979), Southwood (1978), Unwin (1980), Unwin & Corbet (1991)

Reading on hoverflies

Eggs: Chandler (1968a)
Larvae: Coe (1942), Dixon (1960), Dolezil (1970), Doucette and others (1942), Goeldlin (1974), Hartley (1961), Maier (1982), Rotheray (1989b, 1993), Rotheray & Gilbert (1989, 1993), Schneider (1969)
Adults: Collett & Land (1975a,b, 1978) Holloway (1976), Maier & Waldbauer (1979), Morse (1981), Parmenter (1956)
General: Chinery (1976), Colyer & Hammond (1968), Stubbs & Chandler (1978), Stubbs & Falk (1983)

Full reference list

Aubert, J., Aubert, J.-J. & Goeldlin, P. (1976). Twelve years of systematic trapping of syrphids at Bretolet Pass in the Alps. (In French.) *Mitteilungen der schweizerischen entomologischen Gesellschaft*, 49, 115-142.
Begon, M. (1979). *Investigating Animal Abundance. Capture-Recapture for Biologists*. London: Edward Arnold .
Bhatia, M.L. (1939). Biology, morphology and anatomy of aphidophagous syrphid larvae. *Parasitology*, 31, 78-129.
Blackman, R. L. (1974). *Aphids*. Aylesbury: Ginn & Co.
Brower, J.Z. & Brower, L.P. (1965). Experimental studies of mimicry. VIII Further investigations of honeybees *(Apis mellifera)* and their dronefly mimics *(Eristalis* spp.). *American Naturalist*, 99, 173-187.
Buckton, G.B. (1895). *The Natural History of Eristalis tenax or the Drone Fly*. London: Macmillan.
Carthy, J.D. (1979). *The Study of Behaviour*, 2nd edn. Studies in Biology no. 3. London: Edward Arnold .
Chalmers, N. & Parker, P. (1989). *The OU Project Guide*. Field Studies Occasional Publications no. 9.
Chandler, A.E.F. (1966). Some aspects of host-plant selection in aphidophagous Syrphidae. In *Ecology of Aphidophagous Insects*, ed. I. Hodek. Proceedings of a symposium held in Liblice, near Prague, pp. 113-115. Prague, Czechoslovak Academy of Sciences .

Chandler, A.E.F. (1968a). A preliminary key to the eggs of some of the commoner aphidophagous Syrphidae (Diptera) occurring in Britain. *Transactions of the Royal Entomological Society of London*, 120, 199-218.

Chandler, A.E.F. (1968b). Some factors influencing the occurrence and site of oviposition by aphidophagous Syrphidae (Diptera). *Annals of Applied Biology*, 61, 435-446.

Chinery, M. (1976). *A Field Guide to the Insects of Britain and Europe*, 2nd edn. London: Collins.

Cloudsley-Thompson, J. L. (1967). *Microecology*. Studies in Biology no. 6. London: Edward Arnold.

Coe, R. L. (1941). *Callicera rufa* Schummel; colour variation of abdominal hairs in the adult, with a note regarding the longevity of the larva. *Entomologist*, 74, 131-2.

Coe, R. L. (1942). *Rhingia campestris* Meigen (Dipt., Syrphidae): an account of its life-history and descriptions of the early stages. *Entomologist's Monthly Magazine*, 78, 121-30.

Coe, R. L. (1953). Diptera, Family Syrphidae. *Royal Entomological Society Handbooks for the Identification of British Insects*, 10,1-98. (Out of print.)

Collett, T. S. & Land, M. F. (1975a). Visual spatial memory in a hoverfly. *Journal of Comparative Physiology*, 100, 59-84.

Collett, T. S. & Land, M. F. (1975b). Visual control of flight behaviour in the hoverfly *Syritta pipiens*. *Journal of Comparative Physiology*, 99, 1-66.

Collett, T. S. & Land, M. F. (1978). How hoverflies compute interception courses. *Journal of Comparative Physiology*, 125, 191-204.

Colyer, C. N. & Hammond, C. O. (1968). *Flies of the British Isles* (2nd edition). London: Warne.

Cornelius, M. & Barlow, C. A. (1980). Effect of aphid consumption by larvae on development and reproductive efficiency of a flower fly, *Syrphus corollae* (Diptera, Syrphidae). *Canadian Entomologist*, 112, 989-92.

Davies, N. B. (1977). Prey selection and the search strategy of the Spotted Flycatcher *(Muscicapa striata)*: a field study on optimal foraging. *Animal Behaviour*, 25, 1016-33.

Davis, B. N. K. (1991). *Insects on Nettles*. Naturalists' Handbooks 1. Slough: The Richmond Publishing Co. Ltd.

Disney, R.H.L., Erzinclioglu, Y.Z., Henshaw, D.J., Howse, D., Unwin, D.M., Withers, P. & Woods, A. (1982). Collecting methods and the adequacy of attempted faunal surveys, with reference to the Diptera. *Field Studies*, 5, 607-21.

Dittrich, W., Gilbert, F., Green, P., McGregor, P. & Grewcock, D. (1993). Imperfect mimicry: a pigeon's perspective. *Proceedings of the Royal Society of London* B 251,195-200.

Dixon, A. F. G. (1973). *Biology of Aphids*. Studies in Biology no. 44. London: Edward Arnold.

Dixon, T. J. (1960). Key to and descriptions of the third instar larvae of some species of Syrphidae (Diptera) occurring in Britain. *Transactions of the Royal Entomological Society of London*, 112, 345-79.

Dolezil, Z. (1970). Developmental stages of the tribe Eristalini (Diptera, Syrphidae). *Acta Entomologica Bohemoslavica*, 69, 339-50.

Doucette, C. F., Latta, R., Martin, C. H., Schopp, R. & Eide, P. M. (1942). Biology of the Narcissus Bulb Fly in the Pacific North West. *Technical Bulletins of the United States Department of Agriculture*, 809, 1-67.

Edwards, P. J. & Wratten, S. D. (1980). *Ecology of Insect-Plant Interactions*. Studies in Biology no. 121. London: Edward Arnold.

Eisner, T. (1971). Chemical ecology: on arthropods and how they live as chemists. *Verhandlungen der deutschen zoologischen Gesellschaft*, 65, 123-37.

Elton, C . (1966). *The Pattern of Animal Communities*. London: Chapman & Hall .

Fitton, M. & Rotheray, G. E. (1982). A key to the European genera of diplazontine ichneumon-flies, with notes on the British fauna. *Systematic Entomology*, 7, 311-20.

Gatter, W. & Schmid, U. (1990). The migration of hoverflies at Randecker Maar. (In German). *Spixiana (suppl.)* 15, 1-100

Ghorpade, K. (1981). Insect prey of Syrphidae (Diptera) from India and neighbouring countries: a review and bibliography *Tropical Pest Management*, 27, 62-82.

Gilbert, F. S. (1981). The foraging ecology of hoverflies (Diptera, Syrphidae): morphology of the mouthparts in relation to feeding on nectar and pollen in some common urban species. *Ecological Entomology*, 6, 245-62.

Goeldlin, P. (1974). Contribution to systematic and ecological studies of the Syrphidae (Diptera) in western Switzerland. (In French.) *Mitteilungen der schweizerischen entomologischen Gesellschaft*, 47, 151-252.

Hartley, J. C. (1961). A taxonomic account of the larvae of some British Syrphidae. *Proceedings of the Zoological Society of London*, 136, 505-73.

Heal, J. (1979). Colour patterns of Syrphidae. 1. Genetic variation in the Drone Fly *Eristalis tenax*. *Heredity*, 42, 223-36.

Heinrich, B. (1979). *Bumblebee Economics.* Cambridge, Mass: Harvard University Press.

Heiss, E. M. (1938). A classification of the larvae and puparia of the Syrphidae of Illinois exclusive of aquatic forms. *Illinois Biological Monographs*, 16(4), 1-142.

Holloway, B. (1976). Pollen feeding in hoverflies (Diptera, Syrphidae). *New Zealand Journal of Zoology*, 3, 339-50.

Ibrahim, I.A. & Gad, A.M. (1975). The occurrence of paedogenesis in *Eristalis* larvae (Diptera, Syrphidae). *Journal of Medical Entomology*, 12, 268.

Ilse, D. (1949). Colour discrimination in the Drone Fly *Eristalis tenax. Nature, London*, 163, 255–6.

Imms, A. D. (1971). *Insect Natural History*, 3rd edn. London: Collins.

Kay, Q. O. N. (1976). Preferential pollination of yellow-flowered morphs of *Raphanus raphanistrum* by *Pieris* and *Eristalis spp. Nature, London*, 261, 230-2.

Kendall, D. A. & Stradling, D. J. (1972). Some observations on the overwintering of the dronefly, *Eristalis tenax* (L.) (Syrphidae). *Entomologist*, 105, 229-30.

Krebs,J. R. & Davies, N. B. (1987). *An Introduction to Behavioural Ecology* (2nd edition). Oxford: Blackwell Scientific Publications .

Leir, V. & Barlow, C. A. (1982). Effects of starvation and age on foraging efficiency and speed of consumption by larvae of a flower fly, *Metasyrphus corollae* (Syrphidae). *Canadian Entomologist*, 114, 897-900.

Maier, C. T. (1982). Larval habitats and mate-seeking sites of flower flies (Diptera: Syrphidae, Eristalinae). *Proceedings of the Entomological Society of Washington*, 84, 603-9.

Maier, C. T. & Waldbauer, G. P. (1979). Dual mate-seeking strategies in male syrphid flies (Diptera, Syrphidae). *Annals of the Entomological Society of America*, 72, 54-61.

Morse, D. H. (1981). Interactions among syrphid flies and bumblebees on flowers. *Ecology*, 62, 81-8.

Oldroyd, H. (1958). *Collecting, Preserving, and Studying Insects.* London: Hutchinson. (Reprinted 1970.)

Osten Sacken, C. R. (1894). *On the Oxen-born Bees of the Ancients (Bugonia), and their relation to Eristalis tenax, a two-winged insect.* Heidelberg.

Owen, D. F (1956). A migration of insects at Spurn Point, Yorkshire. *Entomologist's Monthly Magazine*, 92, 43-4.

Owen, J. (1981). Trophic variety and abundance of hoverflies (Diptera, Syrphidae) in an English suburban garden. *Holarctic Ecology*, 4, 221-8.

Owen, J. (1991). *Ecology of a garden.* Cambridge: Cambridge University Press.

Parmenter, L. (1956). On *Syritta pipiens* L. (Syrphidae) and its habits. *Entomologist's Record and Journal of Variation*, 68, 211-14.

Phillipson, J. (1966). *Ecological Energetics.* Studies in Biology no. 1 London: Edward Arnold.

Pickard, R. S. (1975). Relative abundance of syrphid species in a nest of the wasp *Ectemnius cavifrons*, compared with that of the surrounding habitat. *Entomophaga*, 20, 143-51.

Proctor, M. & Yeo, P. F. (1973). *The Pollination of Flowers.* London: Collins .

Redfern, M. (1983). *Insects and thistles.* Naturalists' Handbooks 4. Cambridge: Cambridge University Press

Redfern, M. & Askew, R. R. (1992). *Plant Galls.* Naturalists' Handbooks 17. Slough: The Richmond Publishing Co. Ltd.

Ricklefs, R. E. (1991). *Ecology.* 3rd edn. London: Thomas Nelson.

Roberts, M. J. (1970). The structure of the mouthparts of syrphid larvae (Diptera) in relation to feeding habits. *Acta zoologica*, 51, 43-65.

Rotheray, G. E. (1979). The biology and host-seeking behaviour of a cynipoid parasite of aphidophagous syrphid larvae. *Ecological Entomology*, 4, 75-82.

Rotheray, G. E. (1981a). Host searching and oviposition behaviour of some parasitoids of aphidophagous Syrphidae. *Ecological Entomology*, 6, 79-87.

Rotheray, G. E. (1981b). Emergence from the host puparium by *Diplazon pectoratorius* (Gravenhorst) (Hymenoptera, Ichneumonidae), a parasite of aphidophagous syrphid larvae. *Entomologist's Gazette*, 32, 39-41.

Rotheray, G E. (1989a). *Aphid Predators.* Naturalists' Handbooks 11. Slough: The Richmond Publishing Co. Ltd.

Rotheray, G.E. (1989b). Colour, shape and defence in aphidophagous syrphid larvae (Diptera). *Zoological Journal of the Linnean Society* 88, 201-216

Rotheray, G.E. (1993). Colour guide to hoverfly larvae. *Dipterists Digest* (supplement).

Rotheray, G.E. & Gilbert, F.S. (1989). Systematics and phylogeny of European predacious Syrphidae (Diptera) based on larval and puparial stages. *Zoological Journal of the Linnean Society* 95, 29-70

Rotheray, G.E. & Gilbert, F.S. (1993). Systematics and phylogeny of European Syrphidae (Diptera) based upon larval and puparial stages. *Zoological Journal of the Linnean Society*, submitted

Rotheray, G.E. & MacGowan, I. (1990). Re-evaluation of the status of *Callicera rufa* Schummel (Diptera, Syrphidae) in the British Isles. *Entomologist,*109, 35-42.
Růžička, Z. (1975). The effect of various aphids as larval prey on the development of *Metasyrphus corollae* (Diptera, Syrphidae). *Entomophaga*, 20, 393-402.
Schmidt-Nielsen, K. (1975). *Animal Physiology: Adaptation and Environment.* Cambridge: Cambridge University Press.
Schneider, F. (1969). Bionomics and physiology of aphidophagous syrphid larvae. *Annual Review of Entomology*, 14, 103-24.
Scott, E. I. (1939). An account of the developmental stages of some aphidophagous Syrphidae (Diptera) and their parasites (Hymenoptera). *Annals of Applied Biology*, 26, 509-32.
Smith, J.G. (1976). Influence of crop background on natural enemies of aphids on Brussels sprouts. *Annals of Applied Biology*, 83, 15-29.
Smith, K G. V. (1974). Changes in the British Dipterous fauna. In *The Changing Fauna and Flora of Britain*, ed. D. L. Hawksworth, pp. 371-91. London & New York: Academic Press .
Southwood, T. R. E. (1978). *Ecological Methods, with Particular Reference to the Study of Insect Populations.* London: Chapman & Hall.
Spears, E. E. (1983). A direct measure of pollinator effectiveness. *Oecologia, Berlin*, 57, 196-9.
Speight, M. C. D. (1980). The *Chrysogaster* species (Diptera, Syrphidae) known in Great Britain and Ireland. *Entomologist's Record and Journal of Variation*, 92, 145-51.
Stelleman, P. (1978). The possible role of insect visits in pollination of reputedly anemophilous plants, exemplified by *Plantago lanceolata* and syrphid flies. In *The Pollination of Flowers by Insects*, ed. A. J. Richards, Linnean Society Symposium 6, pp. 41-6. London: Academic Press.
Stubbs, A. E. (1980). The rearing of *Cheilosia paganus* and *C. fraterna* (Diptera, Syrphidae). *Entomologist's Record and Journal of Variation*, 15, 1-255.
Stubbs, A. E. & Chandler, P. (1978). A Dipterist's handbook. *Amateur Entomologist*, 15, 1-255.
Stubbs, A. E. & Falk, S.J. (1983). *British Hoverflies: An Illustrated Identification Guide.* London: British Entomological and Natural History Society.
Thornhill, R. & Alcock, J. (1983). *The Evolution of Insect Mating Systems.* Cambridge, Mass.: Harvard University Press.
Unwin, D. M. (1980). *Microclimate Measurement for Ecologists.* London & New York: Academic Press.
Unwin, D. M. & Corbet, S. A. (1991). *Insects, plants and microclimate.* Naturalists' Handbooks 15. Slough: The Richmond Publishing Co. Ltd.
Varley, G. C. (1937). Aquatic insect larvae which obtain their oxygen from the roots of plants. *Proceedings of the Entomological Society of London*, 12A, 55-60.
Verrall, G. H. (1901, reprinted 1969). *British Flies, vol. 8, Platypezidae, Pipunculidae and Syrphidae.* Hampton, Classey.
Wellington, W. G. & Fitzpatrick, S. M. (1981). Territoriality in the Dronefly, *Eristalis tenax* (L.) (Syrphidae). *Canadian Entomologist*, 113, 695-704.
Yeo, P. F. & Corbet, S. A. (1983). *Solitary Wasps.* Naturalists' Handbooks 3. Cambridge: Cambridge University Press.
Zumpt, F. (1964). *Myiasis in Man and Animals in the Old World.* London: Butterworth.

Index

CPSIA information can be obtained
at www.ICGtesting.com
Printed in the USA
BVHW091136140620
581468BV00005B/283

9 781907 807596